Francis Melville Deems

Handbook of Systematic Urinary Analysis, Chemical and

Microscopical

Francis Melville Deems

Handbook of Systematic Urinary Analysis, Chemical and Microscopical

ISBN/EAN: 9783742806123

Manufactured in Europe, USA, Canada, Australia, Japa

Cover: Foto ©berggeist007 / pixelio.de

Manufactured and distributed by brebook publishing software
(www.brebook.com)

Francis Melville Deems

Handbook of Systematic Urinary Analysis, Chemical and Microscopical

HANDBOOK

OF SYSTEMATIC

URINARY ANALYSIS

CHEMICAL AND MICROSCOPICAL.

FOR THE USE OF PHYSICIANS, MEDICAL STUDENTS, AND CLINICAL ASSISTANTS,

BY

FRANK M. DEEMS, M. D., Ph. D.,

FORMERLY LABORATORY INSTRUCTOR IN THE MEDICAL DEPARTMENT OF THE
UNIVERSITY OF NEW YORK; MEMBER OF THE N. Y. COUNTY MEDICAL
SOCIETY; MEMBER OF THE N. Y. MICROSCOPICAL

SECOND EDITION

NEW YORK
THE INDUSTRIAL PUBLICATION COMPANY.
1881.

PREFACE TO THE SECOND EDITION.

Part Second of this little manual was (and is still) published in a separate form. The reception accorded to it by both the medical press and profession was so flattering, that I have been induced to add to it. Owing to the liberality of the publisher, I feel justified in saying that there is no other work on this subject, of a similar size and price, so well and profusely illustrated. In the reading matter, no originality whatever is claimed, and I have often adopted even the very language of others where it seemed to combine accuracy, brevity and clearness. The work is still a compilation. I hope, however, that my intention, as set forth in the preface to the First Edition, will be more completely fulfilled.   F. M. D. ·

Augusta, Ga., October 1, 1881.

PREFACE TO THE FIRST EDITION.

The following plan or method for the systematic examination of the urine, step by step, both chemically and microscopically, is compiled with the intention of supplying students of medicine, clinical assistants, and busy practitioners with a concise guide, which, from its small compass and tabulated arrangement, will serve both as a bedside reference-book and a work-table companion. While it cannot pretend to take the place of larger works upon the highly important subject of urinary analysis, the compiler hopes, from his somewhat extended experience as a teacher of this branch of physical diagnosis, that it will serve to lessen the difficulties in the way of the beginner, and save time to the busy practitioner in his routine examinations.

429 West Twenty-Second Street,   F. M. D.
  New York, October 1, 1880.

LIST OF ILLUSTRATIONS.

FIG.

Frontispiece, Urinary Analysis Set (after Prof. Draper).   PAGE
1. Urinometer and Cylinder, ... 20
2. Separate Crystals of Diabetic Sugar, ... 33
3. Normal Deposit from Ammoniacal Urine, ... 36
4. Extraneous Substances found in Urine. . . . 39
5. Nitrate of Urea, ... 40
6. Oxalate of Urea, ... 41
7. Apparatus for the Quantitative Estimation of Urea, . . 45
8. Common Forms of Uric Acid, ... 49
9. Rarer Forms of Uric Acid, ... 50
10. "Hodge-hog" Crystals of Urate of Sodium, . . . 53
11. Crystals and Amorphous Deposit of Urate of Sodium, and Spherules of Urate of Ammonium, ... 54
12. Spherules of Ammonium Urate, ... 55
13. Hippuric Acid from Human Urine, ... 56
14. Oxalate of Lime, ... 59
15. Dumb-bells and Ovoids of Oxalate of Lime, . . . 60
16. Crystals of ("Stellar") Phosphate of Lime, . . . 63
17. Different Forms of Ammonio-magnesian ("triple") Phosphate, 64
18. Crystals ("fern-leaf") of Triple Phosphate, . . . 65
19. Leucin Spheres and Tyrosin Needles, ... 68
20. Cystin, Hexagonal Tablets and Prisms, . . . 70
21. Xanthin from Human Urine, ... 71
22. Mucus Cloud, ... 73
23. Renal, Vesical, Urethral and Vaginal Epithelium, . 75
24. Mode of Formation of Tube-casts, . . . 80
25. Epithelial and Opaque Granular Casts, . . 81
26. Fatty and Blood Casts, Free Fatty Molecules, . . 82
27. Waxy Casts, ... 83
28. Mucus Casts, ... 85
29. Blood Corpuscles in the Urine, . . . 88
30. Pus Corpuscles, ... 92
31. Human Spermatozoids, ... 94
32. Body and Upper Part of the Tail of a Spermatozoon, . 95
33. Bacteria and Vibriones, ... 96
34. Penicillium Glaucum, Aerial Fructification, . . 98
35. Torula Cerevisiæ, Aerial Fructification, . . 100
36. Sarcina in the Urine, ... 100
37. Apparatus for Collecting Urinary Sediments, . . 118
38. The New Working Microscope, ... 121

HANDBOOK OF
URINARY ANALYSIS.

CHAPTER I.

COMPOSITION OF THE URINE; PHYSIOLOGICAL AND PATHO-
LOGICAL ALTERATIONS; TRANSPARENCY; CONSISTENCE;
ODOR; COLOR; REACTION; SPECIFIC GRAVITY; THE
URINOMETER; QUANTITY.

Healthy urine is a clear, watery, pale yellow, or amber-
colored, fluorescent fluid, of an acid reaction, saline taste,
characteristic (sui generis) odor, and, when passed in the
average quantity of about 1500 c.c., has a mean specific
gravity of about 1020°. It is free from any deposit or sediment, but generally holds in suspension a little mucus and
epithelium.

Composition.

What may be called the average composition of human
urine is shown in the subjoined table. The quantities in
which these various constituents are present vary considerably, according to the circumstances and influences to which
the person is subjected.

Amounts of the Several Urinary Constituents Passed in Twenty-four Hours. (*After Parkes*).

	By an Average Man of 66 Kilos.		Per 1 Kilo. of Body Weight.
Water	1500·000	Grammes	23·0000
Total Solids	72·000	"	1·1000
Urea	33·180	"	·5000
Uric Acid	·555	"	·0084
Hippuric Acid	·400	"	·0060
Kreatinin	·910	"	·0140
Pigments and other substances	10·000	"	·1510
Sulphuric Acid	2·012	"	·0305
Phosphoric Acid	3·164	"	·0480
Chlorine	7·000—(8·21)		·1260
Ammonia	·770		
Potassium	2·500		
Sodium	11·090		
Calcium	·260		
Magnesium	·207		

From the above we see that the greatest amounts are represented by Urea and the Chlorides (chiefly of Sodium). The gases which can be extracted from the urine by the mercurial pump are carbon dioxide (60—150 c.c. in 1000 c.c. of urine). Nitrogen is present in very small quantity, and of oxygen traces only can be discovered.

The Renal Function.

There is a difference of opinion among physiologists concerning the excretion of urine. Bowman holds that the epithelial cells lining the secreting tubules are secretion organs, and that water only is excreted by the tufts, or malpighian bodies, which washes the constituents of the urine out of the epithelial cells. This is the secretory theory. Ludwig assumes that the excretion of urine is a process of filtration taking place in the glomeruli, and a process of diffusion throughout the course of the tubules; the epithelial cells lining the tubules are not taken into consideration at all. This is the mechanical theory. But experimental evidence justifies the conception which the structure of the kidney would lead us *a priori* to adopt—namely, that the secretion of urine by the kidney is *a double process ;* a combination of secretion and filtration. The filtration part of the process is directly dependent on blood pressure. However, there is no question as to the chief function of the kidney being that of an excreting organ, its office being to get rid of substances produced by the activity of other tissues. Its principal work is not to form, but to eject, and the nitrogenized excrementitious matters thus ejected are, in great part, the effete products of tissue-metamorphosis (metabolism), and hence represent the physiological wear of the organism.

Physiological and Pathological Variations.

The preceding table gives the average composition of normal urine. A morbid condition of the system may greatly alter its composition; there may be an absence, a diminution, or an excess of one or more of its physiological elements, or some new and unnatural substance, such as, for example, *albumen, sugar, blood, pus, bile, or casts* may be mixed with it. Owing to these physiological and pathological variations, of both the physical properties and chemical composition of the urine (which fluctuate in proportion as the processes of life fluctuate), its condition affords a clue to the nature of deranged or diseased states of the system, an index to their prognosis, and a guide to their successful treatment (Harley). " At the bedside there is no more important subject of inquiry than the chemical composition of the urine, since from this we obtain the most certain data upon which to ground a diagnosis or furnish a prognosis " (Ziemssen's Cyclopædia).

Every physician should be as thoroughly familiar with the causes and significance of these urinary changes, and the means of detecting them, as he is with cardiac or pulmonary derangements, and their symptoms and physical signs. The more important of these factors influencing the composition of the urine will be pointed out in the following detailed summary of its physical and chemical characteristics.

Transparency.

Normal urine deposits, after standing at rest for some time, a slight cloud (nubecula), derived from the bladder and urinary tract, consisting mainly of mucus and epithelial cells, but in all other respects it is perfectly transparent. *A visible sediment, other than the above mentioned, appearing in the urine within eight to twelve hours is abnormal.* On the other hand, because a given specimen of urine is transparent, it by no means follows that it is therefore normal. Sometimes, on cooling, urine that was passed perfectly clear may become turbid from a precipitation of the urates (of soda, potash, lime, and magnesia), which, though highly soluble in water *at the temperature of the body*, are promptly precipitated from a cold urine. This turbidity is seen usually on a winter morning. On warming this turbid urine, the urates will *re-dissolve*, and the liquid resumes its translucency. Another cause of turbidity of the urine, and perhaps a more common one, is the precipitation of the *earthy phosphates* (of lime and magnesia), which require for their solution an *acid* urine. We shall see, when treating of *Reaction*, that during digestion there is always a diminished acidity, or even an alkalescence, of the urine; and it is at such times, therefore, that this precipitation is most likely to occur. Heat increases this deposit of earthy phosphates, an acid causes its prompt disappearance, being the reverse of the case of the urates. Should the urine be turbid *when voided*, it is a mark of disease, and *pus* is the most frequent cause of this appearance. Opacity, produced by pus or albuminous cellular elements, is *increased* by heat and acids, owing to the precipitation of the albumen, which is an invariable constituent of *liquor puris* and *liquor sanguinis.*

Consistency.

Normal urine is always perfectly fluid, like water, readily flowing through and dropping from a tube of exceedingly small calibre. But in disease, especially in *catarrh of the bladder*, and in retention of urine, the ammoniacal products of the decomposition of the urea often render the pus so thick, viscid, and glutinous that it is difficult or even impossible to pour it from one vessel into another. Such a state of viscidity is called "ropiness." Chylous urine (rare, and peculiar to certain tropical countries) also gives to the urine such an increased consistence as to form, sometimes a thick, firm, jelly-like mass. The froth on normal urine readily disappears; but if the froth be *permanent*, the presence of *sugar*, *albumen*, or the constituents of the *bile*, may be suspected.

Odor.

When freshly-passed, normal urine exhales a peculiar aromatic odor, which gives place, on cooling, to an odor characteristic to itself, and known as urinous. It is said to be due to certain volatile organic acids. The color and odor of the urine are usually modified by the same *physiological* conditions. The addition of a mineral acid greatly intensifies and modifies the urinous odor. When urine loses its natural odor, and becomes fetid and ammoniacal, the change is due to the decomposition of *urea* into *ammonium carbonate* (a highly alkaline salt), and to the formation of sulphur compounds. Should the organic matters be increased, and undergo decomposition at the same time, the odor is at once putrescent and ammoniacal. In cases of destructive renal or cystic disease, and in paraplegia, these alterations begin very quickly after the urine has been voided. A fixed alkali gives to the urine a faint mawkish or an aromatic odor; diabetes is said to impart a sweetish whey-like odor, which changes to that of sour milk on fermenting. Purulent and sanious discharges, a stale, offensive odor, like that of tainted flesh. It must not be forgotten that certain articles of diet (as asparagus, cauliflower, garlic, onions, etc.), and certain drugs (as turpentine, valerian, assafœtida, castor, copaiba, cubebs, sandal-wood oil, etc.) communicate certain peculiar odors to the urine. It is well, therefore, to be aware of what the patient has been taking.

Color.

Healthy urine may be said to be, in general terms, of a pale wine-yellow, or amber color; but even within the limits of health there may be considerable variation, from the palest straw to a yellowish-brown. These variations are due to the relative proportion of the coloring matters to the water. The larger the quantity of urine daily excreted the paler the tint; the less the amount the higher the color. Very *pale* urine is voided by persons *in health*, after copious drinking, especially in cold weather; very *high-colored* urine after prolonged and violent muscular exercise and severe sweating.

At present the whole subject of urinary pigments, both in health and disease, is very imperfectly understood. Whether the natural yellow color of urine be due to a single pigment, the *urochrome* of Thudichum, or to more than one, and what is the exact nature of these pigments, must be left for a while undecided. Urine frequently contains *urobilin*, and there are reasons for thinking that the urinary and bile-pigments are derivatives of *hæmoglobin*. If this be the case an immense number of blood corpuscles must be destroyed daily (and replaced by new ones) in order to give rise to the amount of urinary and bile pigment discharged daily from the body (Foster). Urobilin is found in acute febrile diseases, and points to an increased waste of red blood corpuscles.

The color of the urine varies greatly in disease; it may be pale yellow, green, red, brown, blue, and black. These tints may be due to an excess or diminution of the *normal* coloring matters, to an increase or decrease of the quantity of water holding them in solution, to the presence of blood and bile pigments, and to other *abnormal* coloring agents, and, lastly, to certain articles of *diet* or *medicine* which impart coloration to the urine.

Pale urine is passed by persons suffering from anæmia and chlorosis, diabetes, nervous disorders, such as hysteria and convulsions (urina spastica), and in convalescence from acute diseases. As a rule, *pale* urine indicates *the absence of pyrexia.* However, a pale or normal-colored urine is not invariably an indication of health, as it may contain an abnormal quantity of *urohæmatin*, determined by proper investigation.

High-colored urine is found in fevers, and most acute diseases in which considerable tissue metamorphosis takes place. A dark-colored urine should be examined for excess of pigment; and its presence, when an average amount of urine is passed (1500 c.c.), almost invariably indicates the existence of a more or less grave pathological condition of the system. Reddish-pink urine makes its appearance in various febrile complaints from the addition of an abnormal coloring matter, called, by Prout, *purpurin*, by Heller, *uroërythrin*, and by Harley, *urohæmatin*. It contains iron, and is doubtless a modified hæmatin, being also found especially in diseases where there is evident blood dyscrasia, with destruction of the blood-corpuscles, as in low fevers, septic conditions, etc. Such a high-colored urine will give, when diluted with water, according to the amount of dilution, all the colors more commonly observed in this fluid, thus strengthening the hypothesis that these various tints are only dilutions of one and the same coloring agent.

Certain medicinal and poisonous substances, finding their way into the system, may produce various alterations of color in the urine. Santonin imparts a rich golden-yellow to acid urine, and an orange-red to alkaline urine; gamboge, senna, and picrotoxin, yellow color; rhubarb, a deep gamboge-yellow, changed to red by the addition of ammonia; turpentine, violet; gallic and tannic acids, dusky; creosote and arseniuretted hydrogen, a black discoloration. Among other substances than the above-named, aloes, blackberries, raspberries, indigo, logwood, madder, etc., color the urine. Strong coffee darkens the urine. It is of some importance, therefore, to know what a patient has been eating or drinking. When urine containing bile is kept for several days, it sometimes changes from a brown to a grass-green color, owing to the oxidation of the biliary pigments. *Blue urine* is most frequently met with in cholera and typhus fever, and is due to the presence of *indican*. *Black* urine has been seen in those suffering from melanotic cancer. All of these abnormal pigments have an intense affinity for *uric acid* and the *urates*, and hence color them, when they are thrown out as a deposit.

Fluorescence.

Normal urine is sometimes markedly fluorescent; as yet we are unable to state the substances that produce it. Alkaline urine, by *reflected* light, appears greenish; by transmitted light yellowish-red (Hoffman and Ultzmann).

Reaction.

Healthy human urine of the twenty-four hours is acid, the amount of this acidity being equivalent to from 2 to 4 grammes (30–60 grains) of oxalic acid. It is due to the presence of *acid sodium phosphate* ($PO_4 H_2 NA$). There is no property of the urine of more varied and important significance than its reaction, for therewith is intimately connected the occurrence of several kinds of urinary deposits. The reaction of the urine is liable to be affected by food, the cold bath, medicinal substances, general diseases, and the decomposition of the secretion.

(*a*) *Food.*—The reaction of the urine holds a close relation to the digestion of food. It may be neutral, or even alkaline, shortly after taking food, while the process of digestion is going on, again becoming more and more acid up to the time of the next meal. The alkaline urine, voided after food, owes its reaction to *fixed* alkali, and not to ammonia. The more remote consequence of a meal, however, is to maintain, and even increase, the acidity. Animal has this effect more than vegetable food. As a rule, if the urine of a person, living on a mixed diet, becomes alkaline in less than twenty-four hours after being passed, there exists some disease, either of the general system or of the urinary organs, which demands immediate attention.

(*b*) *The Cold Bath.*—The urine is said to become invariably alkaline after prolonged immersion of the body in a bath at a lower temperature than that of the body.

(*c*) *Effects of Medicines.*—Both mineral and vegetable *acids*, when administered in large quantities, tend somewhat to raise the acidity of the urine; but their effect is inconsiderable. Urine that is habitually alkaline cannot be rendered acid by the internal administration of even large quantities of these substances. *Alkaline substances* have a much more powerful influence, and it is an easy matter to render the urine strongly alkaline at pleasure. This effect may be attained by the caustic and carbonated alkalies, or by the alkaline salts of certain groups of vegetable acids—acetic, tartaric, citric, malic, and lactic acids. The least disturbing to the digestive organs are the bicarbonates of potash and soda, and the acetates and citrates of the same bases. It requires from 300 to 400 grains of the bicarb. potass., and about as much of the acetate and citrate, given in divided doses during the twenty-four hours, to keep the urine steadily alkaline in the adult.

(*d*) *General Disease.*—In persons of debilitated constitutions, in the anæmic state which *follows* subacute rheumatism and gout, in chlorosis, atonic dyspepsia, chronic vomiting, and even in chronic phthisis, the urine may present the character of alkalescence (from fixed alkali). The clinical significance of urine, rendered alkaline by the *fixed* alkalies, is not serious, as this condition is an occasional accompaniment of debility, from whatever cause it may be due. Persons who pass such an alkaline urine are generally suitable subjects for a tonic and stimulating treatment, and exercise in the open air.

Changes Undergone by Urine on Standing.—After urine has been discharged, the natural acidity *increases* for some time, owing to the formation of fresh acid, apparently by some kind of fermentation. This increase of acid frequently causes a precipitation of *urates*, which the previous acidity has been insufficient to throw down. After a while, however, the acid reaction gives way to alkalinity. This is caused by the decomposition of urea $(NH_2)_2CO$, into ammonium carbonate $(NH_4)_2CO_3$), a strongly alkaline salt. This change is accomplished through the agency of a specific ferment. This ferment, as a general rule, does not make its appearance except in urine exposed to the air; it is only in unhealthy conditions that the alkaline fermentation takes place *within the bladder*. The mode by which urea is decomposed into ammonium carbonate is thus explained. The urea takes up water, and, by a simple rearrangement of their particles, becomes converted into ammonium carbonate:

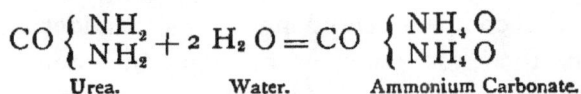

$$CO \left\{ \begin{array}{l} NH_2 \\ NH_2 \end{array} \right. + 2\,H_2O = CO \left\{ \begin{array}{l} NH_4O \\ NH_4O \end{array} \right.$$

Urea.　　　　　Water.　　　Ammonium Carbonate.

This transformation is brought about with great rapidity by contact with *any decomposing organic matter*—e.g., mucus, pus, blood, epithelium, albumen, etc.; also when the urine is very dilute and feebly acid or alkaline. The peculiar odor so characteristic of stale urine is chiefly due to the ammonium carbonate thus produced.

Significance of the Alkaline Reaction.—Urine which is alkaline from *fixed* alkali, is always *secreted alkaline by the kidneys;* it deposits, if at all, simple amorphous phosphate of lime, of which the particles have no tendency to accrete into gravel or calculi; it has a sweet aromatic odor; is perfectly bland and innocuous to the mucus membranes, and *is not associated with inflammation of the urinary passages.*

Ammoniacal urine, on the other hand, is only in the rarest instances, and under the gravest circumstances, *secreted ammoniacal* by the kidneys; but it usually becomes so by an after-change in the lower urinary passages, or after it has been voided. It is always sedimentary, depositing a mixture of the amorphous phosphate of lime and crystals of the ammonio-magnesian, or triple, phosphates, sometimes with the addition of lumpy spheres and rude dumb-bells of urate of ammonium, which have a strong tendency to aggregate into masses. This urine has an ammoniacal, and often an offensive putrescent odor, and is highly irritating to the mucus membranes, exciting inflammation of them if the contact be long continued.

A urine alkaline from fixed alkali (potash or soda) reflects a state of the blood; a urine alkaline from ammonia (if alkaline when voided) generally points to a local affection of some part of the mucus membrane of the lower urinary passages—generally that of the bladder. Any condition which interferes with the complete emptying of the bladder in micturition favors the production of ammoniacal urine. Paraplegia, with paralysis of the bladder, obstinate urethral stricture, enlarged prostate, calculous concretions, morbid growths, or foreign bodies in the bladder, are all sooner or later implicated with ammoniacal urine. This ammoniacal urine irritates the mucus membrane, and induces cystitis; and the purulent secretion thus engendered reacts on the urine and favors its decomposition. Thus the two conditions mutually aggravate each other, and perpetuate each other's existence after the original cause has passed away. Cystitis may, in this way, persist for years after the removal of a stone, or the cure of a stricture, which was its original cause.

Specific Gravity.

The specific gravity of healthy urine may range from 1000·3° (after copious drinking on an empty stomach) to 1030° (after food), pure water at 60° F. being 1000°. The *usual* range of density of the normal mixed urine of the twenty-four hours is from 1015° to 1025°—average 1020°. It is never lighter than water. The physiological variations in the specific gravity of the healthy secretion depend upon the relative quantity of solids it contains, and it is to judge of the solid matter it contains that we take the specific gravity. The specific gravity varies not only with the time of the day, the constitution of the individual, the food and drink, rest, and amount of exercise taken, but upon numerous other causes, such as, for instance, the state of the skin and lungs.

The urine that is passed immediately after drinking (urina potus) is pale-colored, faintly acid, and of a low specific gravity (1002° to 1015°). That passed in the morning, after sleep (urina sanguinus), is darker in color, of a more acid re-action, and a higher specific gravity (1015° to 1020°), while that voided some hours after eating (urina cibi, or chyli), although having a still higher specific gravity than the last (1020° to 1030°), is neither so dark in color nor so acid in reaction as the morning urine. Seeing that the normal physiological variations of density exhibit these very consider-able ranges, caution must therefore be exercised in drawing inferences from any unusual depression or elevation of the density in disease. If, however, the urine exhibits *habitually*, and especially the morning urine (urina sanguinis), a density *below* 1015°, the presence of *albumen* may be suspected; if the density persists at a still lower point (1005° to 1008°) the existence of diabetes insipidus. After hysterical paroxysms the urine may be abundant in quantity, and of an exceedingly low density (urina spastica). On the other hand, a density above 1025°, especially in a pale, *apparently* dilute urine, is strongly suspicious of the presence of *sugar ;* and the higher densities, from 1035° to 1050°, belong almost entirely to *diabetes mellitus.* Roberts says, the heaviest urine ever sub-mitted to him had a density of 1065°, and did not contain a particle of sugar, but a very large quantity of albumen! A high density urine, free from sugar, indicates concentration, and more particularly a large percentage, or excess, of *urea.* In the febrile state there is an absolute increase of urea, uric acid, and the sulphates, with a diminished elimination of water; consequently the specific gravity ranges high (1030° to 1038°). Whenever there is rapid wasting of the tissues, especially with concurrent sweating, or diarrhœa, the urine has a high density. On the other hand, when the specific gravity is abnormally low (less than 1015°), we may, with great safety, suspect the presence of some exhausting non-inflammatory complaint, such as, for example, Bright's disease.

A high-density urine with a pale color, and a low-density with a deep tint, are equally signs of disease, as are also those cases in which, with small volume, there is low specific gravity, or, with great volume, high specific gravity.

The Urinometer.

The simplest way of estimating the specific gravity of the urine is by means of the urinometer (Fig. 1). Although not

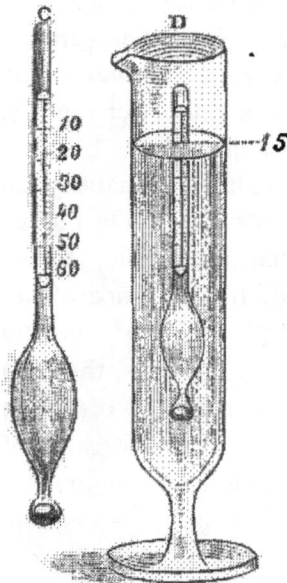

Fig. 1.—URINOMETER AND CYLINDER.

strictly accurate, it is sufficiently so for practical purposes. This instrument consists of a glass float, weighted with a bulb of mercury below, and having above a graduated stem. As the specific gravity of a liquid increases as the temperature is lowered, and decreases as it is raised, these instruments are constructed for use *at a certain temperature*—60° or 62° F; hence, the urine to be tested should either be artificially brought to a temperature of 60° F. (15·54° C.), when attempting to take its specific gravity, or corrections for temperature may be made from the following table:

Bouchardat's Table of Corrections of Specific Gravity for Temperature.

CENTIGRADE SCALE.

NON-SACHARINE URINE.				SACCHARINE URINE.			
Subtract.		Add.		Subtract.		Add.	
0°	·09	15°	0·0	0°	1·3	15°	0·0
1°	·09	16°	0·1	1°	1·3	16°	0·2
2°	·09	17°	0·2	2°	1·3	17°	0·4
3°	·09	18°	0·3	3°	1·3	18°	0·6
4°	·09	19°	0·5	4°	1·3	19°	0·8
5°	·09	20°	0·9	5°	1·3	20°	1·0
6°	·08	21°	0·9	6°	1·2	21°	1·2
7°	·08	22°	1·1	7°	1·1	22°	1·4
8°	·07	23°	1·3	8°	1·0	23°	1·6
9°	·06	24°	1·5	9°	0·9	24°	1·9
10°	·05	25°	1·7	10°	0·8	25°	2·2
11°	·04	26°	2·1	11°	0·7	26°	2·5
12°	·03	27°	2·3	12°	0·6	27°	2·8
13°	·02	28°	2·5	13°	0·4	28°	3·1
14°	·01	29°	2·7	14°	0·2	29°	3·4
15-60°F.	0·0	30°	3·0	15°	0·0	30°	3·7

For Fahrenheit Table, see page 103.

As many of the urinometers sold by dealers are utterly worthless, every instrument, before being used, should be tested with distilled water at 60° F. (15·54° C.), into which it should sink to the mark 0°, or 1000°.

Method of Taking Specific Gravity.—The cylindrical vessel, usually supplied with the urinometer, should be filled about three-fourths with urine, *holding the cylinder obliquely while filling, in order to avoid foam,* which greatly interferes with accurate reading. If this forms, get rid of it either by blotting paper, or overflowing and then partially decanting. The urinometer must readily move up and down the cylinder. If the cylinder be too small, the capillary attraction between its walls and the urinometer will prevent the latter from sinking as low as it should. The stem of the instrument must also be kept from touching the walls of the cylinder. Allow it to *gradually* sink until it will no longer descend spontaneously. Since the fluid accumulates around the stem of the urinometer from capillary attraction, the specific gravity appears to be lower than it really is, when read while the eye is *above* the level of the surface; hence, to obtain a correct reading, the eye must be lowered to the lower level of the surface of the fluid, and the number on the stem cut by the *lower convex edge of the fluid* read off, and noted. Then gently depress the urinometer, about *one degree only,* no more (or its stem, becoming covered with urine, will be rendered heavier, and produce an error in the reading). Having done this, allow it to rise again by removing the finger. When it has come to rest, the number may be again read; the second estimation is made to correct any mistake that may have taken place on the first reading. If the quantity of urine to be examined is too small to fill the cylinder enough to float the urinometer, it may be diluted with a quantity of *distilled* water, sufficient to fill the cylinder to the required height. From the specific gravity of this mixture may be calculated that of the urine, by multiplying the number above 1000 by the total number of volumes of the mixed fluid. Thus, suppose it is necessary to add four times as much water as urine, to enable us to use the urinometer, that is, to make five volumes, and the sp. gr.

of the *mixed* fluid is 1004°, then that of the urine would be $1000° + (4 \times 5) = 1020°$.

The knowledge of the specific gravity of a few ounces of urine is a matter of little value. To render the observation in any way useful, the whole quantity passed in the twenty-four hours must be collected and mixed, and the specific gravity of a portion of this taken. This remark applies equally to all *quantitative* analyses.

Determination of the Amount of Solids from the Specific *Gravity.*—An approximate determination of the amount of solids existing in the mixed urine of 24 hours may be made in a few minutes, and with sufficient accuracy for ordinary clinical purposes, by using Trapp's Formula 2, or Haeser's 2·33. Thus, having determined the correct sp. gr. of the urine, multiply the *two last figures* of the number expressing this specific gravity by 2 (for urine below 1018°), by 2·33 (for urine above 1018°), and the result gives the amount of solids, in grammes, existing in 1000 c.c. of the urine. For instance, a person passes 1250 c.c. of urine in 24 hours, of sp. gr. 1020. The last two figures (20) being multiplied by 2·33, gives 46·6, which is the amount in grammes of solid matters contained in 1000 c.c. of the urine. But if 1250 c.c. were passed in the twenty-four hours, the calculation required consists in simply multiplying the whole amount of urine passed in the 24 hours (1250) by the amount of solids in the 1000 c.c. of the urine (46·6), and then dividing this by 1000; thus:

$$\frac{1250 \times 46\cdot6}{1000} = 58\cdot25 \text{ grammes.}$$

and so with any amount of urine in 24 hours, whether it exceeds or falls short of 1000 c.c.

The quantity of solids secreted in the urine in twenty-four hours varies from 55 to 75 grammes.

Quantity of the Urine.

Closely connected with the specific gravity, and holding an inverse relation to it, is the total quantity of urine discharged in the twenty-four hours. Vogel estimates the diurnal average to be 57 ounces; Becquerel 44 ounces; Hoffman and Ultzman, and other recent observers, about 50 ounces. Dr. Parkes found the average daily quantity of urine in healthy male adults, between twenty and forty years of age, to be fifty-two and a half fluid ounces (52½ fl. oz.)` The extremes were thirty-five and eighty-one ounces (35 and 81). Assuming the usual quantity in the male to be about fifty (50) ounces, it may be stated in general terms (Flint), that the range of normal urine is between thirty and sixty (30 and 60); and that, when the quantity varies much from these figures, it is probably due to some pathological condition.

All that has been said of color and specific gravity in this respect is true of the quantity of urine, though in an inverse ratio. It is necessary that one should be acquainted with the physiological as well as pathological conditions on which the quantity and quality of the urine depends. These are (*a*) the drink, (*b*) the kind of food, (*c*) the state of the cutaneous and pulmonary exhalations, (*e*) the length of time the urine is retained in the bladder, (*f*) the condition of the stools, (*g*) the sex, (*h*) the age of the individual, (*i*) the influence of remedies, and (*k*) sleep.

1st. *The Influence of Drink upon the Amount of the Urine.*— The flow of urine is essentially regulated by the quantity of fluids drank; controlled, however, in a most important degree by the pulmonary and cutaneous exhalation, and by the call of the system for water at the time. The more active the skin, *cæteris paribus*, the less water is excreted by the kidneys. Whenever the quantity of drink is diminished, the amount of urine is correspondingly decreased. When the blood and tissues contain their full complement of water, any further potation results in immediate diuresis, whereby the superabundance is carried off. But when the organs and tissues of the body are craving for more water, a large quantity may be drank without causing diuresis. The kidneys eliminate water in strict accordance with these conditions—it being an essential and important part of their function to regulate the aqueousness of the blood. They are able to separate water at an almost unlimited rate—equally, at least, to the capacity of the gastric vessels to absorb it.

2nd. *The Influence of Food upon the Amount of Urine.*—All the urinary ingredients increase after meals. This influence of food over the daily quantity of urine, both liquid and solid, seems to depend chiefly on one of its elements, viz., the *nitrogen.* The more nitrogenized the diet, the greater the quantity of urine excreted. The influence that diet exerts over the relative amount of solids excreted by the kidneys is immense. They, like the water, are increased by an animal (nitrogenized), diminished by a vegetable, and still further reduced by a non-nitrogenized diet. (100 parts of animal food contain, on an average, 15 per cent. of nitrogen, whereas the same amount of vegetable food contains only 2 to 5 per cent.) The same rule holds good regarding the effect of drink on the solids; for, just as the amount of solids taken into the stomach influences the quantity of water eliminated by the kidneys, so the amount of water taken into the stomach augments the amount of solids excreted by the kidneys in the twenty-four hours. This may be caused partly by the excess of water in the circulation accelerating to some extent the metamorphosis of the tissues, and partly, no doubt, by the fact that by increasing the quantity of fluid in the intestines more food is dissolved, and consequently more solid matter is absorbed

into the circulation, the excess of which, beyond the requirements of the system, is excreted by the kidneys.

3rd. *Influence of the Cutaneous and Pulmonary Exhalations.*—The more rapid the pulmonary and cutaneous exhalations, the less the amount of urine excreted. Hence, we invariably observe that, after much exercise, the urine is scanty, dark-
colored, and of high specific gravity. It is sometimes, indeed,
so concentrated, after profuse perspiration, that it irritates the
urethra. Normal sweat contains uric acid, urea, phosphates,
and chlorides; and in disease even the insoluble oxalate of
lime may be excreted in the perspiration in such quantity as
to cover the skin with a white crystalline crust. Cutaneous
exhalation may, therefore, for a time replace the renal function. The influence of the pulmonary exhalation is nearly of
equal importance. Much more urine is passed in winter than
in summer; on a cold and damp day than on a warm and
dry one. Expired air has been found to contain chloride of
sodium, urate of soda, urea, and urate of ammonium.

4th. The quantity of urine passed also depends, to a great
extent, upon the time it is retained in the bladder. If detained in the bladder, the water, and even many of the solids,
may be reabsorbed into the circulation.

5th. *Influence of the Stools.*—The greater the quantity of
liquid passed from the bowels, the less is ejected from the
bladder; so that, in cases of diarrhœa and dysentery, patients
void much less urine than in a state of health.

6th. *Influence of Sex and Age.*—As a rule, males pass more
urine than females (men 35–64, women 26–45 fluid ounces).
Children, for their size, pass proportionally more water than
adults. As age advances, the absolute, as well as relative
proportion of urine, diminishes. The amount of water passed
by the kidneys appears, therefore, to be proportional to the
metamorphosis of the tissues

7th. *The Influence of Disease and Remedies.*—Disease greatly affects the quantity of the urine; in some few it is increased, as, for example, in polydipsia and diabetes; also in the middle stages of atrophic degeneration of the kidneys. Temporary excess occurs after hysterical paroxysms and certain other convulsive attacks in both males and females. Any increase of tension in the arterial system, as in some cases of hypertrophy of the left ventricle of the heart, gives rise to an increased secretion of the urine. Remedies have also a powerful effect on the amount of urine excreted by the kidneys. Diuretics, such as, for example, spiritus aetheris nitrosi, increase it; while others have just the contrary effect. Mineral substances—such as iron and copper—lessen the quantity of the urine, while cantharides and arsenic almost totally arrest its secretion.

The urine is always scanty in cirrhosis of the liver; in some forms of Bright's disease through their entire course, and in the last stages of all forms; in any condition of the heart which directly or indirectly causes passive congestion of the renal veins, whereby the circulation through the kidneys is impeded. It is very scanty, sometimes approaching or reaching total suppression in the early stages of acute Bright's disease. In the scarlatinal form of nephritis, and in yellow fever, the urine is frequently suppressed for twenty or thirty-six hours. All types of violent fever and inflammation are liable to be complicated with suppression of the urine. The same may be said of the collapse period of cholera. Partial or total suppression also occurs in the later stages of all organic diseases of the kidneys; and when any mechanical obstacle obstructs the flow of the urine. It has followed catheterism, even in old cases of stricture, when instruments had been repeatedly used before without any ill effects.

8th. *Influence of Sleep.*—The mean *hourly* discharge of ·

solid urine during the waking hours is about 33·14 grains, while the average of the hours of sleep is about 15·33 grains, or less than one-half.

From what has been said before, and owing to the fact that there are three kinds of urine passed during the twenty-four hours—the urina sanguinis (morning), the urina potus (after drinking), and the urina cibi (of digestion)—and that these differ in specific gravity, owing to the variable amounts of solids they contain; and in reaction, it is easy to see the absolute necessity of collecting all the urine discharged by the patient during twenty-four hours, and analyzing a sample taken from the whole amount. In all cases where it is desired to make a *quantitative* analysis, we must be aware of the absolute quantity.

CHAPTER II.

ABNORMAL SUBSTANCES IN THE URINE; ALBUMEN, SUGAR, EXTRACTIVES, BILE.

Albuminuria.

Albumen—$C_{216} H_{169} N_{27} S_3 O_{68}$.

Albumen is not found in any proportion in healthy urine, but is the most common and the most important of the abnormal ingredients found in the urine in disease, especially in diseases of the kidneys. Its presence in the urine is due, however, to *so many conditions, that the mere fact of its presence yields* little direct information; but when correctly interpreted, it furnishes a key to several grave pathological states, which would otherwise remain in great obscurity. Albumen may be present only as incidental to the presence of some other fluid in the urine, such as *blood, pus, leucorrhœal discharge, or spermatic fluid.* The following conditions, among others, may give rise to its slight and temporary presence in the urine, the excessive use of a diet chiefly or exclusively albuminous (eggs), and dyspepsia. Setting aside unimportant exceptions, albuminuria must always be looked upon as a grave symptom of disease; and when discovered, its signification becomes an anxious question to the practitioner. The pathological states in which albumen appears constantly or occasionally in the urine, may be arranged into the following groups:

1. Acute and chronic Bright's disease of the kidneys; structural changes.

2. Pregnancy and the puerperal state.

3. Febrile and inflammatory diseases (zymotic diseases, such as scarlet fever, measles, small-pox, typhoid, cholera, yellow fever, ague, diphtheria, etc.; inflammatory diseases, such as pneumonia, peritonitis, traumatic fever, articular rheumatism, etc.)

4. Impediments to the circulation of the blood (emphysema, heart disease, abdominal tumours, cirrhosis, etc.)

5. A hydræmic and dissolved state of the blood, with atony of the tissues, (purpura, scurvy, pyæmia, hospital gangrene); also in hæmaturia.

6. Saturnine, arseniuretted hydrogen, and carbonic dioxide poisoning.

When albumen is found in the urine, the important point to decide is, whether it indicates the existence of *organic* disease of the kidneys or not. *This cannot be positively done without the aid of the microscope.* The greater the quantity of albumen, the more likely is the existence of renal diseases. Of all urines, a pale, dilute, abundant albuminous urine is the most surely indicative of Bright's disease. Contrariwise, as a rule, with very few exceptions, a slightly albuminous, and at the same time dense and high-colored urine, is more indicative of pyrexia, or some impediment to the circulation of the blood, than of Bright's disease. When the urine is found *permanently* albuminous, and there exists neither pyrexia nor thoracic diseases, or other recognizable condition which can account for the albumen, the inference is almost irresistible that there exists a primary organic renal disease. If, in addition to this, there be a very abundant deposit, containing *casts* and much *renal epithelium,* or numerous casts and cells in a state of fatty degeneration, the proof is almost positive.

The least indicative of primary renal diseases of serious import are blood-casts, and very small transparent casts in scanty numbers. *The search for renal casts must always follow the detection of albumen in the urine.* A *low* specific gravity aids in distinguishing the albuminuria consequent upon structural changes in the kidneys from the numerous other forms of albuminuria, since in what is commonly called Bright's disease the specific gravity ranges between 1005° and 1014°, the usual average being 1010° and 1012°. In the other forms of albuminuria the specific gravity has no limit, ranging from 1005° to 1035°. The reason for the low specific gravity of Bright's disease is that the disorganized renal tubes are incapable of eliminating the urea and other urinary crystalloids. The lower, therefore, the specific gravity in these affections, the more dangerous is the case.

Sugar.

Sugar $(C^6 H^{12} O^6) + H^2 O$ *in the Urine.* (Glycosuria; diabetes mellitus).

Whatever may be said as to the existence or non-existence of sugar in healthy urine, it seems to be certainly established that normal urines, and the vast majority of morbid urines, do not contain sugar in sufficient quantity to be detected by the most delicate direct testing; that when sugar is present in sufficient quantity to be of clinical importance, it can be detected with certainty by direct testing; and, lastly, when such direct testing reveals the *constant* presence of sugar, it is invariably a grave pathological sign. *Small quantities* of sugar may appear in the urine for very short periods, being a temporary and incidental consequence of some physiological or pathological antecedent, such as, for example, the eating of an excessive quantity of saccharine or amylaceous food;

after the administration of chloroform, ether, chloral, turpentine, nitro-benzole, nitrite of amyl, etc.; during recovery from cholera; after a paroxysm of whooping-cough, asthma, or epilepsy. These are cases of incidental glycosuria. Glycosuria has been found to follow upon injury of the head, with or without fracture of the skull; clot in the pons varolii; softening of the base of the brain; abscess and tumours of the cerebellum; bony spiculæ in the falx; disease of the sympathetic nerve; tumour of the pneumogastric nerve; excessive brain work; intense grief; sudden mental shock; spontaneous gangrene; blow in the epigastrium; pregnancy; uterine disease; anthrax; disordered digestion; exposure to cold; hereditary influence, etc. In fact, the presence of sugar in the urine, like the discoloration of the skin in jaundice, is not of itself the disease, but merely the most prominent sign of several widely-differing abnormal conditions. In the other class of cases, the glycosuria is a more intense, *permanent*, uninterrupted symptom, and is associated with a serious departure from health. The term *diabetes* is applicable alone to this latter class, and it is only in true diabetes that sugar occurs in *large* quantities. Moreover, the excess of water in diabetes is enormous. Since, however, as has just been stated, the urine may become temporarily saccharine under certain conditions, quite apart from genuine diabetes, care must be exercised not to conclude too rashly from the mere finding of a little sugar in the urine that this formidable disease exists. In addition to the presence of a *considerable quantity* of sugar in the urine, it must be ascertained, by oft-repeated examinations, that its appearance is *persistent* and uninterrupted, and that there is a less or greater increase in the volume of the urine.

Microscopical Appearance.—Crystals of diabetic sugar may be prepared by simply evaporating a few drops of diabetic urine to dryness on a glass slide, provided the urine is *very rich in sugar*, and contains but *little urea and other salts.* Fig. 2 represents crystals of grape sugar, obtained by the

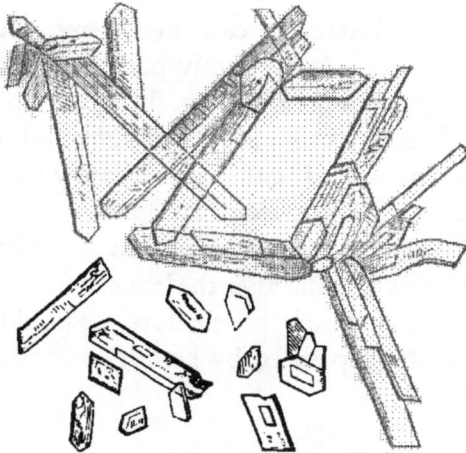

Fig. 2.—SEPARATE CRYSTALS OF DIABETIC SUGAR × 130. (Beale.)

above-mentioned process. The most characteristic form of the crystals is that of the rhomboidal prism, occasionally arranged in arborescent tufts. They form very beautiful objects when examined by polarized light. The sugar may also crystallize in warty concrements that consist of cauliflower leaflets.

Extractive Matters.

The presence in the urine of the blood-extractives, indicates merely the escape of blood material, and proves the existence of congestion or inflammation of some part of the urinary surfaces. In Bright's disease, according to Rees, the extractives can be found in the urine before albumen is met with, and also that they exist after the albumen has disappeared: thus on the one hand warning us of the approach of

albuminuria, and, on the other, against too early a belief .n
convalescence; for, as he justly observes, so long as the blood
is losing its extractives, so long is the patient in peril.   The
presence of the extractives also enables us to diagnosticate
nephritic irritation from renal calculus *before* albumen, blood,
or pus has appeared.   It is highly probable that extractives
will be found preceding albumen in urine in most cases
(Da Costa).   Healthy urine is scarcely affected by tincture of
galls, but the tincture of galls immediately precipitate the
blood extractives.   This precipitate must not be confounded
with that of the earthy and potassium salts, which are thrown
down from all kinds of urine after the lapse of five or ten minutes,
by the spirit contained in the tincture.   Should *albumen* be
present, it must be separated by boiling and filtration before
applying the test.

Bile in the Urine.

The bright golden-red color of normal human bile is due
to the presence of *bilirubin* ($C^{16} H^{16} N^2 O^3$).   It occurs sometimes abundantly in the urine of jaundiced persons.   It imparts to the urine a dark-brown or greenish-yellow color.   All
the constituents of the bile may appear in the urine, or only
the pigment, without the acids or their salts.   The more permanent and marked occurrence of the bile pigment in the
urine is always attended with jaundice, but it is sometimes
found transiently, and in small quantities, without yellowness
of the skin.   It may be detected both before and after the
discoloration of the skin.   The biliary acids are not of necessity present in the urine of icterus.   Nor is the cause of
their presence clearly understood.   Probably so great a quantity of them is produced that they cannot undergo the normal
changes in the blood, and are therefore met with in the urine.

CHAPTER III.

URINARY DEPOSITS, PRODUCED BY CHANGES IN THE URINE
ON KEEPING; EXTRANEOUS SUBSTANCES FOUND IN THE
URINE; DIVISION; DETAILED DESCRIPTION OF UNORGAN-
IZED URINARY DEPOSITS; UREA (NITRATE AND OXALATE);
URIC ACID; HIPPURIC ACID; AMORPHOUS URATES; SODIUM
URATE (CRYSTALLINE); AMMONIUM URATE (CRYSTALLINE);
OXALATE OF LIME; CALCIUM PHOSPHATE (AMORPHOUS AND
CRYSTALLINE); AMMONIO-MAGNESIUM (OR TRIPLE) PHOS-
PHATE; LEUCIN AND TYROSIN; CYSTIN; XANTHIN.

Urinary Deposits.

Preliminary Remarks.—The terms urinary sediment and
urinary deposit are given to the various substances precipitated from the urine, in which they are held either in solution
or suspension, when this liquid is kept at rest for a longer or
shorter time.

Healthy acid urine, when freshly passed, shows no sediment, except the very light scanty flocculent deposit of mucus
described on p. 72. Under a variety of abnormal circumstances more abundant sediments make their appearance in
the urine. A knowledge of these circumstances is highly important, and prevents erroneous conclusions.

Change in Reaction.—Healthy urine exposed to the air
undergoes a regular series of spontaneous changes. The first

is a progressive increase of the acid reaction (called by Scherer *acid urinary fermentation*). As a consequence of this, there appears usually, first, a precipitation of the *amorphous urates*, then of *uric acid*, often associated with *oxalate of lime*, and spores of the *torula cerevisiæ* (yeast fungus), or of *penicilium glaucum* (mould fungus). The acidity goes on steadily increasing for 4 or 5 days (sometimes longer), and afterwards

Fig. 3.—THE NORMAL DEPOSIT FROM AMMONIACAL URINE.

Showing crystals of ammonio-magnesian phosphate, amorphous phosphate of lime, and spheres of urate of ammonium. (Roberts.)

begins to decline as the urine passes into a state of decomposition. It then becomes turbid or opaque, from the development of myriads of minute linear particles (*bacteria, vibrios*). The *alkaline fermentation* has now set in. The urea becomes converted into ammonium carbonate, which gives a highly alkaline reaction to the urine, and a change takes place in the character of the sediment. The *amorphous urate* becomes changed into dark round masses of *ammonium urate* (which is the only urate that can exist in *alkaline* urine). The *uric acid* crystals give place to bright prisms of *triple phosphate* and an abundant sediment of *amorphous calcium phosphate* sinks to the bottom of the vessel. The confervoid vegetations cease to grow with the change of reaction, and finally perish as the secretion becomes putrid. The student is advised to practice upon specimens of his own urine, which are allowed to go through the various stages alluded to above. Urines of low specific gravity, or of feeble acidity, either do not pass through this series of changes at all, or do so in a very imperfect manner. Their acidity undergoes no appreciable increase, and in a day or two, or in *even a few hours*, especially in the summer, they become ammoniacal. This transformation is brought about with great rapidity by contact with any decomposing organic matter, especially by contact with decomposed urine.

The physical and chemical characters of the urine have now become so altered by these changes that *it is unfit for clinical examination,* and should *invariably be rejected,* except in cases where the transformation takes place *within* the urinary passages, and a more natural specimen is therefore not procurable. *In consequence of these changes, it is desirable to examine the urine within as short a time after its emission as possible, certainly within a few hours.* Certain organic deposits are liable to be greatly altered, or altogether destroyed, by an exposure of 12 to 24 hours, even when the more obvious characters have not undergone a perceptible change. Blood corpuscles, renal epithelium, and casts, soon putrefy and are rapidly disintegrated. Pus, pavement epithelium, and spermatozoids resist much longer without effacement of their microscopical characters; and the last may generally be recognized without difficulty in urine far advanced in putrefaction.

Extraneous Substances Found in the Urine.

The following are some of the most important of the extraneous matters likely to fall under the observer's notice when the urine is examined microscopically: Human hair, cat's hair, worsted, wool, cotton and flax fibres, splinters of wood, portions of feathers, scales of the moth, fibres of silk, milk, oily matter, potato, wheat, and rice starch, tea leaves, bread crumbs, chalk and sand. In several cases of vesico-rectal fistula the author has met with the *débris* of the food in the urine. These and other bodies may fall into into the urine accidentally; they may come from filthy receiving vessels, or they may have been put into the urine for the express purpose of deceiving the practitioner. The importance of recognizing them is obvious, and decided advantage will be derived from subjecting many of these substances to microscopical examination, so that when met with *in the urine* their nature may be at once recognized. Some of them will be found represented in Figure 4.

Having now glanced at the causation of a large proportion of urinary deposits, a detailed consideration of them follows in order.

Urea.

Urea $(NH_2)_2CO$.

Urea is the chief constituent of normal urine. When pure it crystallizes from a *concentrated* solution in the form of long, thin, four-sided, glittering needles, or prisms, which are colorless and odorless. Owing to its exceeding solubility in water (being, in fact, deliquescent), it *never occurs as a spontaneous urinary deposit.* It possesses a bitter cooling taste. When heated to above 120° C., the crystals first liquefy, then burn, giving off ammonia, and leave no residue. Heated

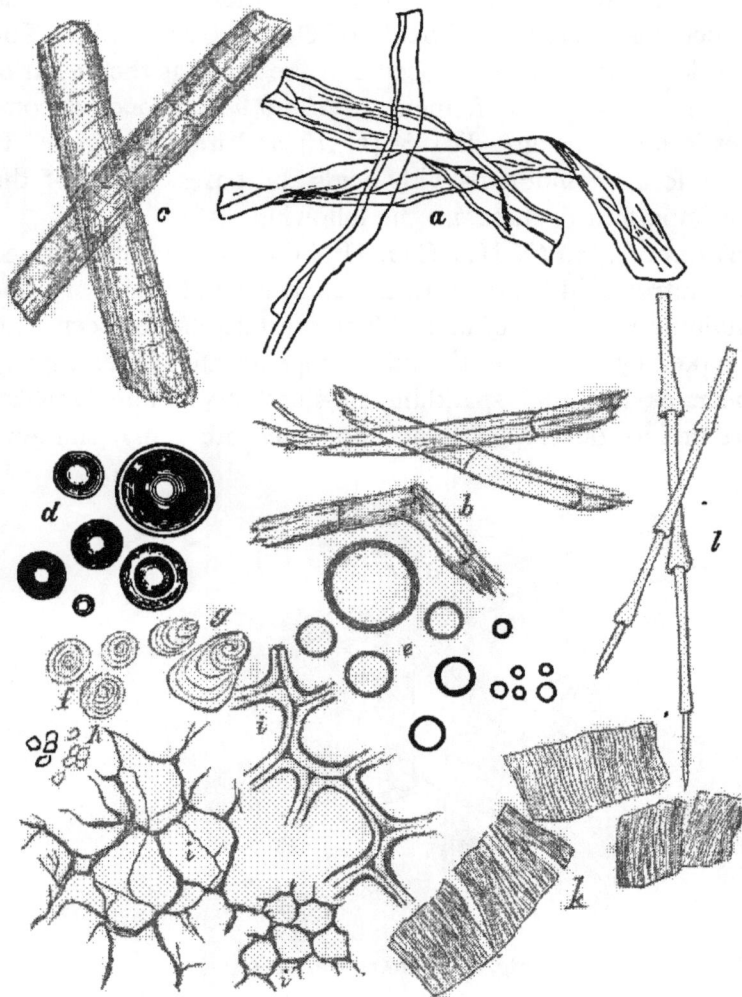

Fig. 4.—EXTRANEOUS MATTERS FOUND IN URINE.

a. Cotton fibres.   b. Flax fibres.   c. Hairs.   d. Air bubbles.
e. Oil globules.   f. Wheat starch.   g. Potato starch.   h. Rice
starch granules.   i, t, i. Vegetable tissue.   k. Muscular fibres.
l. Feathers.

with strong acids or alkalies, decomposition ensues, the final products being carbonic dioxide (CO_2) and ammonia. The same decomposition also occurs in the urine as the result of the action of a specific ferment. *Nitrous acid* at once decomposes it into carbonic dioxide (CO_2) and free nitrogen. It readily forms compounds with acids and bases. Two of the most important of these are the following:

Nitrate of Urea ($(NH_2)_2CO.HNO_3$).—This salt is formed when nitric acid is added to the concentrated urine, or urine containing an excess of urea. The crystals may be seen with the naked eye. Under the microscope nitrate of urea usually appears as beautiful, sparkling, six-sided crystalline lamellæ, more or less developed, lying upon each other like shingles.

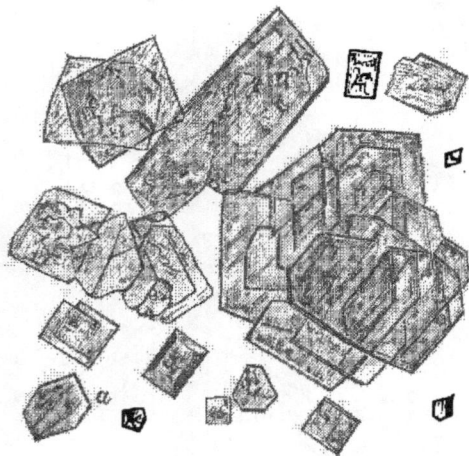

Fig. 5.—NITRATE OF UREA.

Obtained by adding nitric acid in moderate quantity to slightly concentrated urine in a test-tube, and allow it to crystallize slowly. $\times 130$. (Beale.)

The character of the crystals varies slightly with the amount of acid added, the degree of concentration of the urine, and the slowness or rapidity of the process of crystallization.

When *slowly* crystallized, nitrate of urea forms fine prisms. In albuminuria they may take the form of brush-shaped needles.

Oxalate of Urea $(N H_2)_2 C O)_2$. $H_2 C_2 O_2 + H_2 O$.—If, instead of nitric acid, a *concentrated* solution of *oxalic acid* be added to the concentrated urine, numerous crystals of the *oxalate of urea* will be formed. It, like the nitrate, crystallizes in rhomboidal or hexagonal plates and prisms (Fig. 6), some-

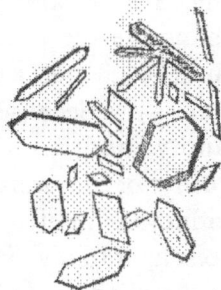

Fig. 6.—OXALATE OF UREA.
Obtained by adding oxalic acid to concentrated urine. $\times 215$.

times separate, sometimes adhering together in groups. The crystals are more perfectly formed than those of the nitrate, and the inclination of the angles is different.

In addition to the microscopic appearances of urea crystals (obtained on evaporating the urine), both the nitrate and oxalate should be formed and examined.

From a physiological and clinical point of view, urea must be regarded as the most important constituent of the urine, since it is the most constant and occurs in the greatest amount. Its source seems to be two-fold; firstly, from the disintegration of the tissues, and, secondly, from the excess of nitrogenized food absorbed into the circulation. The average amount excreted daily by a healthy adult man has been estimated to be at the rate of

3½ grains per pound of the weight of his body, or about 385 to 500 grains in the 24 hours (30 + 40 grammes). But the amount varies considerably from various causes, such as age, sex, diet, exercise, disease, remedies, and individual peculiarities. Before the age of fifteen or eighteen years, the amount of urea (and other solid matters) excreted by the kidneys, in proportion to the weight of the body, is much greater than in the adult. In women the normal excretion is generally less than in men. It has been ascertained that a purely animal diet will increase the amount of the urea fully 2-5ths ; a vegetable diet will diminish it ⅓ ; and a non-nitrogenized diet will reduce it more than ½. Because it is formed by the system in greater quantity, and accumulates in the blood more rapidly than any other of the urinary constituents, when its elimination by the kidneys is interrupted, and because it is a powerful irritant poison, rapidly inducing convulsions, coma and death, it can be readily understood how exceedingly important it is to estimate the quantity of urea eliminated in certain diseases. Urea is not the only, but the most important, measure of tissue change. Moreover, as the production of urea is profoundly affected by the quality of food taken, it sometimes becomes a matter of prime importance to reduce its production by regulating the diet. If we wish to diminish the urea by means of diet, it will be necessary to administer arrowroot, sago, tapioca, and other such amylaceous foods, well sweetened with sugar; and when a more nourishing diet than this is demanded, cream, cod-liver oil, or any other fatty matters will have an equally good effect in reducing the amount of urea. On the other hand, when it is deemed advisable to increase it, animal soups, eggs, milk, jellies, and other nitrogenized substances, together with a fair portion of common salt, and coffee *without sugar*, will answer the purpose.

The amount of urea formed is *increased* by nitrogenized food, by copious water drinking, and common salt, cubebs, cantharides, atropine; in all febrile affections (except yellow fever, in which it is diminished); the exanthemata; inflammatory affections, such as, for example, pneumonia and meningitis; epilepsy; and markedly in diabetes, both saccharine and insipid.

It is *decreased* by non-nitrogenized food—tea, coffee, sugar, starch, fat; citrate of iron and quinine (Harley); digitalis, colchicum, acetate and phosphate of soda; in paralysis, cholera, Bright's disease, and before the paroxysms of gout and asthma.

There are several quantitative tests for urea; Liebig's, Davy's and others. We shall give only one; namely, the simple but accurate process, recently introduced by Drs. Russell and West, and adopted by Roberts and others. In this test, as in all the others, it is necessary to ascertain first the presence of albumen, and if it be present, to separate it by the following method: A measured quantity of urine, 100 or 200 c.c., is heated in a Berlin dish, the bottom of which is protected by a piece of iron wire gauze, over a small naked flame, to nearly the boiling point. If the albumen do not separate in flakes, dilute acetic acid is very carefully added until the acid reaction be marked. If too much acetic acid be added, part of the albumen may be redissolved. The heat must be moderate, and only just rise to the boiling point, or the urea will be decomposed. The urine, after cooling, is then to be filtered into the same measure that was used at the beginning to ascertain the volume (100 or 200 c.c.), the dish and filter washed with small quantities of distilled water, which are added to the filtrate, until the urine stands at exactly 100 or 200 c.c., according to the initial quantity. This fluid is then ready to be used in the estimation of the urea (chlorides, phosphates, sugar, etc.)

If the urine be *free from albumen*, the following method may be used without the above preparation of the urine. It is based upon the fact that when urea is acted on by an alkaline hypobromite, nitrogen is given off, and this nitrogen, when collected and measured, gives an estimation of the urea decomposed.

The hypobromite solution used in this process is made by dissolving 100 grammes of common solid caustic soda in 250 c.c. of water and then gradually adding 25 c.c. of bromine. The solution may be had of any manufacturing chemist, but it must be quite freshly prepared. The apparatus necessary is figured in the adjoining woodcut (Fig. 7). It consists of the tube, A, which is about 9 inches long, narrowed somewhat before the closed end be reached, which end is blown out into a bulb, B. The free end is fitted by means of an india-rubber cork into the hole, G, at the bottom of the trough, C D, which stands on three legs. When an estimation is to be made, 5 c.c. of urine are measured off with a pipette, and allowed to run down the side of the tube, A, into the bulb, B. Distilled water is then added by means of a wash-bottle, to wash away the urine adhering to the sides of the tube, but in quantity enough only to fill up the bulb as high as the constriction, or a very little above it. A glass rod tipped with india rubber is then introduced into the tube, A, so that the india rubber fills up the constriction and acts as a cork. Care must be taken that no air-bubbles are present below the constriction. The hypobromite solution is then poured into the tube until it be quite full, and the trough filled with common water. The graduated tube, F, is then filled with water, the thumb slipped over it, so that it contains no air-bubbles, and then inverted in the trough. The glass rod tipped with india rubber is withdrawn, and the graduated tube, F, immediately slipped over the mouth of the tube, A. The mouth of this should project well up into the tube, F, so as to prevent the escape of any of the gas. The reaction

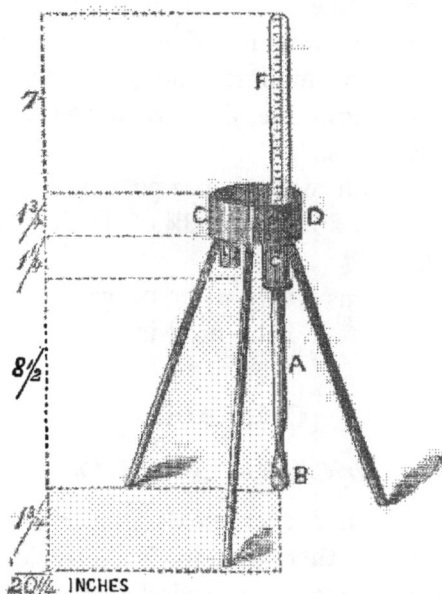

Fig. 7.—APPARATUS FOR THE QUANTITATIVE ESTIMATION OF UREA.*

begins immediately, but in order to bring it to an end as quickly as possible, the bulb should be heated by a spirit lamp or Bunsen's burner till the bulb be hot to the hand.   In five minutes the amount of gas may be read off.   After some hours the gas is lessened in quantity; it is thus important to read it off as soon as the reaction is over.   The amount of gas measured by the divisions on the tube gives the percentage of the urea in the specimen of urine examined.

The estimation of the total amount of urea excreted in the

*The description and cut of the apparatus for the estimation of urea are taken from Dr. Wickham Legg's excellent little work on the urine.

24 hours is now easy   All that is to be done is to multiply the total number of cubic centimetres of urine passed by the figures read off on the measured tube.   Thus, if the amount of urine passed in the twenty-four hours be 1770 c.c., and the percentage of the urine 1·8, it is only necessary to multiply 1770 by 1·8 to get the amount of urea in milligrammes.

This will be 31,860 milligrammes; and since 1000 milligrammes equal one gramme, the total amount of urea passed will be 31·86 grammes.

If the urine contain more than 2 per cent. of urea, the urine must be diluted with an equal bulk of distilled water, and the results doubled.

This method seems likely to be of great value in clinical reasearch, as it may readily be used in the wards of a hospital.

Uric Acid.

Uric Acid.—$C_5 H_4 N_4 O_3$.

Uric acid seldom, if ever, exists in a free state in normal urine; its presence, therefore, in freshly-voided urine in an uncombined state, must be regarded as a pathological condition.   It occurs sparingly in normal human urine.   The quantity excreted daily by a healthy man ranges from 5 to 15 grains, average 6 to 9 grains.   It is, however, the chief constituent of the urine of carnivora, and of the excrement of birds, reptiles, and insects.   Like urea, its source appears to be two-fold; first, from the disintegration of the nitrogenized constituents of the body; and, secondly, from the transformation of the excess of albuminoid food.   By some it is considered as a less oxidized stage of urea.   Whatever accelerates oxidation increases the amount of *urea* eliminated, and diminishes that of the uric acid; while whatever retards oxidation decreases the urea and increases the uric acid.   Usually, however, urea and uric acid increase and diminish together. Uric acid is *increased* in all fevers (yellow fever excepted), exanthematous affections, acute inflammatory diseases, particularly in hepatic, cardiac, and splenic; also in skin diseases, such as eczema; in leucocythæmia, and after the paroxysm of gout. In acute rheumatism the little red granules visible to the naked eye form a deposit in the urine soon after it is voided. Port wine and beer increase the amount. Nearly double the quantity is excreted in winter than in summer. It is *diminished* in yellow and remittent fevers; in diabetes; frequently in albuminuria; in cholera, anæmia, chlorosis, hysteria, and before and during the paroxysm of gout. Tea, coffee, atropine, colchicum, quinine, acetate of potash, and cod-liver oil are all said to diminish it.

Clinical Significance of Uric Acid.—This has to do not so much with the variations of its *quantity*, whether absolute or relative, as with *its precipitation in the free state*, and the *time* and *place* of that precipitation. A deposit of uric acid occurring 12 or 20 hours after emission has no pathological significance, for healthy urine usually deposits uric acid as a normal event, in the course of the acid urinary fermentation. If the deposit takes place *within 3 or 4 hours after emission*, it certainly is unnatural; but it does not require special therapeutical attention. If, however, uric acid be precipitated *before the urine cools, or immediately after it*, it shows that the same event may occur within some part of the urinary renal tract, and thus give rise to the formation of gravel and stone, with all their painful and dangerous consequences. Again, the presence of free uric acid is no proof that it is being excreted in excess; the only inference that can be made from this fact is that the urine is extremely acid.

(We may remark here, in general, that any deposit occurring within twelve hours after its discharge is always a sign of something wrong, and although, as we have seen, it may occur from comparatively trivial causes, whenever it does take place without an appreciable cause, in the otherwise apparently healthy excretion, it is a sign not to be disregarded, since under such circumstances it is not unfrequently either the forerunner or associate of gravel or stone. Uric acid either alone or combined with a base, is the commonest ingredient of all renal and vesical calculi [hence the term lithic acid], and of those concretions deposited in the joints in gouty diseases. No period of life is exempt from calculi. Infants have been born with stone in the bladder.)

The union of uric acid with the bases is very feeble. If from any cause, therefore, the urine become excessively acid after its emission, a deposit of uric acid is liable to occur. The addition of a very small quantity of almost any acid is sufficient to decompose the urates, when the uric acid appears, after a few hours, in a crystalline form.

As a urinary sediment, uric acid is always impure, and the crystals possess a *yellowish or brownish* color. This is due to the fact that in crystallizing the uric acid takes up a part of the coloring matter of the urine, just as sugar—a white substance—when crystallized in a solution of carmine, yields red crystals. This is an important fact to keep in mind, for the majority of *crystalline* substances met with in the urine are perfectly *colorless*, even when crystallized in highly-colored urine; hence this proneness on the part of uric acid crystals (and of the urates) to take up the coloring matter of the urine is a great aid in readily distinguishing them from other substances. The higher the color of the urine, the deeper will be the color of the crystals; the paler the urine, the lighter the crystals. In abnormal urines, containing blue, black, or saffron-colored pigment, the uric acid crystals will be blue, black, or yellow. Chemically pure uric acid is a light, white crystalline powder, tasteless and odorless; nearly insoluble in cold water, alcohol and ether; soluble in alkalies and many alkaline salts.

Microscopic Characters.—The primary or typical forms of uric acid are a four-sided rhomb and a six-sided plate, and to some modification of these figures the protean diversities of uric acid crystals may all be referred. The crystals com-

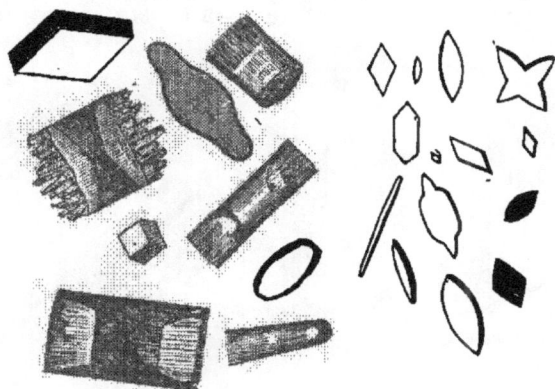

Fig. 8.—COMMON FORMS OF URIC ACID.

monly seen are well marked diamond-shaped plates, rounded lozenges, ovoids, oblongs, squares, comb and barrel-like shapes, cubes, six-sided tablets, rod-stars, spikes, and fan shapes (Fig. 8). Rarer forms, halbert, dumb-bell (Fig. 9).

The dumb-bell form may be compared to a bundle of hay constricted at its middle. The most curious and varied forms are generally found in albuminous urine, frequently crystallizing around tube-casts and masses of epithelium. Sometimes they appear in the form of rows, as if they had been deposited in one of the tubuli uriniferi, which is not at all improbable. The various crystalline forms of uric acid are not always found in a given specimen of urine. A point to remember is that, with some very rare exceptions, uric acid crystals are almost *invariably colored*, yellow, red or brown.

varying from the palest fawn (sand) to the deepest amber, or orange-red (cayenne pepper), and this aids greatly in distinguishing them from other deposits.   They may exist singly, but are very commonly aggregated into rosettes.   The more transparent plates polarize light beautifully, and exhibit the most splendid colors if a piece of selenite be placed beneath the slide containing them.   Whenever any difficulty arises regarding the chemical nature of a crystal, from its having

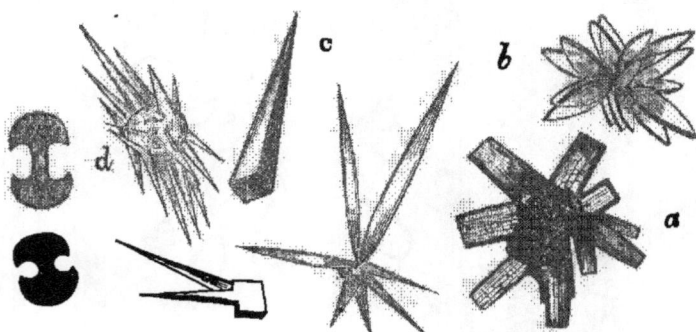

Fig. 9.—RARE FORMS OF URIC ACID.

assumed an unusual form, add to it a drop of *caustic potash*, and, after the crystal has become dissolved, neutralize the solution with an excess of *acetic acid*, and most probably the uric acid will recrystallize in one of its more characteristic or typical forms, but in small *colorless* crystals.   From urine acidulated with hydrochloric acid, large square crystals with two opposite sides smooth and two jagged (the so-called "fine-toothed comb" crystals), are generally deposited.   The form of the crystal is much affected by the strength of the acid which is added;   the weaker the acid the more regular the crystals.

Artificial Production for the Purpose of Study.—Add about a drachm of acetic or hydrochloric acid to a pint of urine, and let it stand from 12 to 24 hours.

Amorphous **Urates.**

(*Syn.* Brick-dust, lateritious deposit.)

We have seen that, although always present in normal urine, uric acid is rarely spontaneously deposited from it *in a free state*, owing to the fact that all the uric acid in normal urine is combined with alkaline bases (soda, potash, ammonia, lime, and magnesia) in the form of very soluble salts. Now, whenever, from any cause, these are in excess, or the water of the urine has been diminished in quantity, they may be deposited in an amorphous state. This deposit is the most common and least important of all the urinary sediments. The amorphous urates have not a fixed and constant composition, but vary considerably in different samples. Sometimes one base and sometimes another predominates. The urates of sodium and the urates of ammonium generally exist in largest quantity; the urates of potassium, lime, and magnesium existing only in minute traces. A conjunction of the following conditions usually determines the precipitation of the amorphous urates, even in healthy persons: A high density (1027°), low temperature, and an acid reaction; and the more acid the urine the more liable is the occurrence of this deposit. A drop of acetic or nitric acid will often cause an immediate precipitation of the amorphous urates from a previously clear urine; and when the urine becomes sedimentary after twelve or twenty-four hours, this is due to the increased acidity produced by the acid urinary fermentation.

Physiologically, a fawn or brownish-colored urate deposit may be expected after profuse sweating, watery evacuations at stool, violent exercise, prolonged abstinence from food and drink, and in cold weather on getting out of bed. Urine containing an excess of urates is never turbid *when freshly passed ;* it is only when the urine has cooled that the peculiar muddiness is observed.   Change of diet, abnormal fatigue, and any nervous shock will also suffice to determine a temporary deposit of this kind.

Pathologically, the commonest determining cause of the precipitation of the amorphous urates is the febrile state ; even a slight degree of pyrexia, as in a common "cold," is usually accompanied by a urate deposit.   Functional digestive derangements are generally attended by pale urate deposits in the urine.   In young children the milky urine, which alarms mothers, is due to a deposit of peculiarly white urates.   The frequent or constant occurrence of a brownish or red deposit, with or without pyrexia, should awaken suspicion of some serious organic disease ; such as disease of the lungs, heart, liver, spleen, or any other part, attended with emaciation and waste of the tissues, is usually accompanied with a deep-colored abundant urate deposit.

The urates generally have a pale fawn-color, but it may vary from snow-white, through every tint, to brick-red, pink, or purple.   Dr. Bird remarks, that in the white, fawn, or brick-red deposit there is a deficient state of the cutaneous functions ; whilst in the pink, crimson, or purple variety there is more or less evidence of functional or organic derangement of the liver, spleen, or other organs influenced by the portal circulation.

Microscopic Appearance.—Under the microscope the amorphous urates are seen as irregular particles or granular powder. They may be mistaken for phosphate of calcium.   The differential test is their behavior with acids ; the phosphate is dissolved by acetic or hydrochloric acid ; the urates are *gradually transformed into crystals of uric acid.*   From carbonate of lime, which also occurs in a granular form, *both* the amorphous urates and the calcium phosphate may be distinguished by the effervescence of the carbonic acid, which occurs when a strong acid is added to a carbonate.

. Crystalline Urates.

Urate of sodium and urate of ammonium are sometimes deposited separately in urine in the *crystalline* form, and under circumstances wholly different from those which determine the precipitation of amorphous urates.

Urate of Sodium, though generally amorphous, occasionally takes a crystalline form. This substance is a constituent of gouty concretions, (so-called chalky deposits). It is a comparatively rare *spontaneous* deposit in the urine. It occurs, occasionally, in gout, and in the febrile state, especially of infants and children, when the urine is excessively scanty, concentrated, and long detained in the bladder. It derives its chief *clinical* importance from the fact that it is precipitated *within the urinary renal passages.* The spiny crystals irritate the mucus membrane of the bladder and urethra, and the

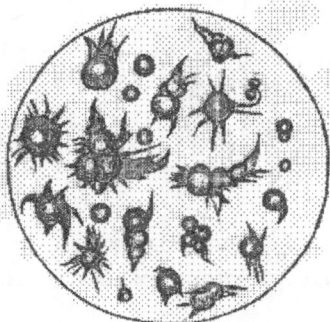

Fig. 10.—HEDGE-HOG CRYSTALS OF URATE OF SODIUM.
Spontaneously deposited from the urine of a child. (Roberts.)

latter canal may even be occluded by impacted masses of this deposit. It may also form a nucleus around which calculous matter may aggregate. The great comparative frequency of vesical culculi in children is not improbably owing to the occurrence of this deposit in the numerous fugitive febrile attacks to which children are subject. (see Fig. 10.)

Under the microscope, a spontaneous deposit of sodium urate exhibits irregular, opaque, globular and lumpy masses, or spherules, from which project spiny crystals, sometimes straight, sometimes variously curved—the so-called hedge-hog crystals.   At the end of the acid, and the beginning of the alkaline fermentation, true *prismatic crystals* of acid sodium urate may occur, arranged in star-like masses.   (See Fig. 11.)

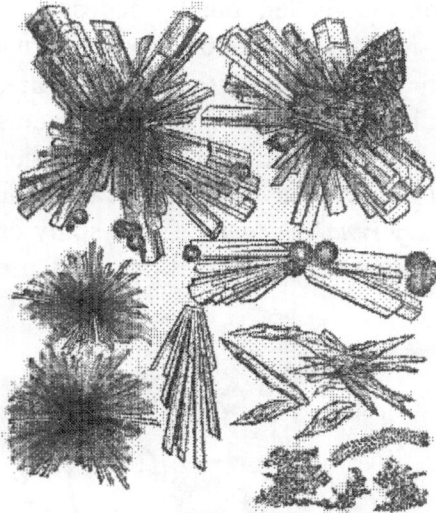

Fig. 11.—CRYSTALS AND AMORPHOUS DEPOSIT OF URATE OF SODIUM.
SPHERULES OF AMMONIUM URATE.   (Ranke.)

The urate of soda from a gouty concretion, under the microscope, is resolved into myriads of long delicate needles, arranged in bundles or stars, or lying separately.   These acicular forms are never deposited spontaneously in the urine; but they may be readily produced by adding a little liquor sodæ to the common amorphous urate, in a watch-glass, and allowing the solution so formed to concentrate by evaporation in the air.

Urate of Ammonium.

When urine becomes strongly alkaline, it is liable to precipitate urate of ammonium, in addition to the mixed phosphates, which are necessarily deposited under those circumstances. It usually has a dense white color; but may possess a beautiful violet hue. This deposit has no special *clinical significance*, its occurrence being merely an incident in the ammoniacal decomposition of urine. It is the only urate found in alkaline urine.

Microscopical Characters.—(See Fig. 12.) Urate of am-

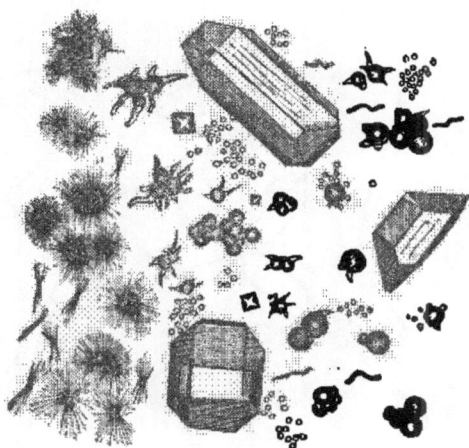

Fig. 12.—SPHERULES OF AMMONIUM URATE.

From alkaline urine, with crystals of oxalate of lime and triple (ammonio-magnesian) phosphate. (Ranke.)

monium is met with in the form of spherules and globular masses, which appear almost black by transmitted light, owing to their opacity. These spherules (thorn-apple spherules) are found in ammoniacal urine, and are the only urates found in alkaline urines. Both this and the urate of sodium respond to the murexide test.

Hippuric Acid.

Hippuric Acid ($C_{18} H_9 NO_6$).

Hippuric acid exists in normal urine in larger quantity than does uric acid, the amount excreted in health varying from (0·5 to 1·0 grammes) 7 to 15 grains per 24 hours. Being, however, an extremely soluble substance in water, it is *rarely*, if ever, to be found as a spontaneous deposit. Some authors think the acidity of normal urine is due to it, it being a much stronger acid than uric acid. It is greatly increased by the eating of much fruit, such as cranberries, blackberries, plums,

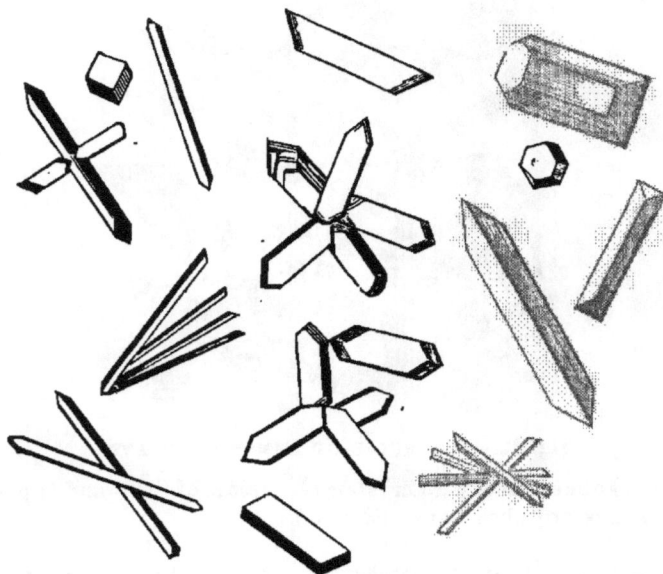

Fig. 13.—HIPPURIC ACID FROM HUMAN URINE. (Harley.)

etc., and also by the ingestion of the balsams of Peru and tolu, and benzoic acid. If 10 or 12 grains of benzoic acid be taken at bedtime, hippuric acid will be found in the urine next morning, if a few drops of hydrochloric acid be added to the urine, then concentrating it to the consistence of a syrup, and allowing it to cool. Yet very little is known regarding the *clinical* significance of hippuric acid. In diabetes it seems, sometimes, to replace uric acid, and is often absent from the urine in jaundice.

Microscopical Characters.—The crystals are, in general, rhombic prisms, columns, or long fine needles. The fine needle-shaped ones, either separate or combined in stellate groups, are the most common. Occasionally they are massed in dove-tail fans, and sometimes they form perfect rosettes. When directly crystallized from human urine they are dark-colored, from absorption of the coloring matter of the urine. From the *triple phosphate* (the only sediment with which they are likely to be confounded) hippuric acid is distinguished by (*a*) its occurrence in *acid* urine only, while triple phosphates are only found in *alkaline*, or at least neutral, liquids; (*b*) a drop of hydrochloric acid dissolves the phosphates and leaves the crystals of hippuric acid intact.

Oxalate of Lime.

(*Oxaluria ; oxalic acid diathesis.*)

After the urates, oxalate of lime is the most common unorganized urinary sediment. Urine depositing it is usually high-colored and acid, and frequently a deposit of uric acid and the amorphous urates are conjoined with the oxalate of lime. Oxalic acid appears to be manufactured even in the human body; for, whenever the metamorphosis of some of the animal products is interrupted, a deposit of oxalate of lime makes its appearance in the urine. It is now generally admitted that urea is the *final* product of the retrograde metamorphosis of the nitrogenized tissues; and also that it is in the form of urea that the greater part of the albuminoid group of foods is excreted from the system. Uric acid is supposed to be one of the *intermediate* products of this metamorphosis, and so also is oxalic acid. Diet has an important influence in its production, since there is no doubt that a large amount of oxalic acid finds its way into the system along with the food and drink. All vegetables and fruit rich in oxalic acid, such as garden-rhubarb, sorrel, tomatoes, etc.—citric, malic, and tartaric acids, often cause a temporary deposition of the oxalate of lime. Carbonated alkalies are sometimes transformed into oxalates in the body.

Clinical Significance.—A slight occasional deposit of oxalate of lime in the urine cannot be said positively to constitute a diseased condition. It is not infrequently found in the urine of typically healthy persons during the acid urinary fermentation. A *constant and large* deposit is proof of an abnormal state, which renders liable the formation of an oxalate of lime ("mulberry") calculus, which is one of the most annoying forms of calculi. Oxaluria is a condition which accompanies the lighter or severer forms of illness, and has its proximate origin in an *impeded metamorphosis*—that is, an insufficient activity of that stage of oxidation which changes oxalic into carbonic acid. The chief, if not the whole, source of oxalic acid, is in the nitrogenized constituents of the food and blood; everything, therefore, which retards the metamorphosis of these constituents, occasions oxaluria. Such retardation may be determined by the following causes: (*a*) abuse of nitrogenized, saccharine, and amylaceous articles of food; (*b*) insufficiency of the red blood corpuscles, and, eventually, diminished oxidation; (*d*) insufficient enjoyment of pure fresh air; (*e*) organic lesions, which in any way impede respiration and the circulation of the blood; (*f*) conditions of the nervous system which bear a character of depression, whether these arise primarily from mental derangement or from pathological states of the blood; (*g*) excess of alkaline bases in the blood. Oxaluria, therefore, arises from a variety of conditions—many of them unaccompanied by appreciable departures from health—in which there is *mal-assimilation of food or imperfect disintegration of the tissues.* The most recent authorities are disposed to attribute little significance to the so-called oxalic acid diathesis (Roberts, Beale, Bencke).

Microscopical Characters.—Oxalate of lime crystals are exceedingly minute objects (less than 1-1000 of an inch in

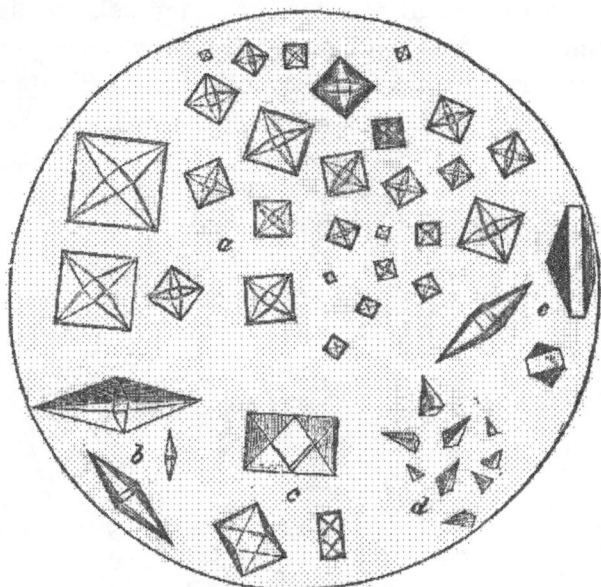

Fig. 14.—OXALATE OF LIME.

a, b, c, Octahedra in various positions; *d,* Pyramids; *e,* Pyramids with intervening square bases. (Roberts.)

diameter, usually), requiring for their detection a high power, else they may be easily passed over as granules of amorphous urates, or spores of fungi. They are best examined with a power of from 400 to 600 diameters. They may be divided

into four well-marked groups: (1) *Octahedra* crystals (which
alone is *characteristic* of calcium oxalate); transparent, or of
an extremely faint greenish tint, with sharply-defined edges
and angles (Fig. 14).   If the light be bright, these octahedral
crystals, seen in their short diameter, resemble a square or
cube, crossed obliquely by two bright lines (the so-called
"envelope" appearance); when very small they appear as a
square with a bright point in the centre; when viewed in the
longer diameter they may be said to be made up of two four-
sided pyramids, placed base to base; they polarize light very
faintly under special conditions.   (2) *Dumb-bells* (very much
more rarely met with) of the typical dumb-bell appearance,

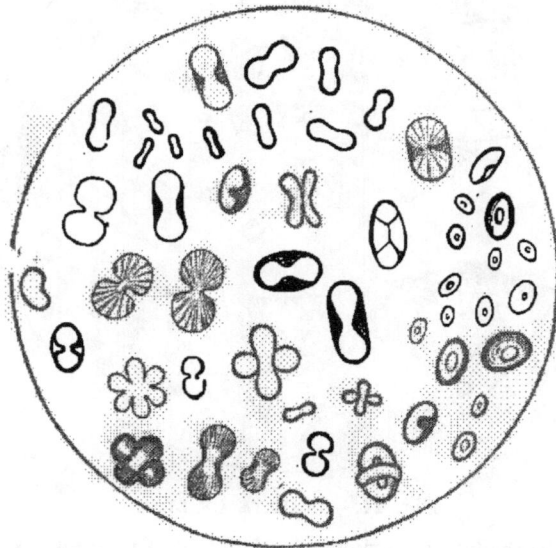

Fig. 15.—DUMB-BELLS AND OVOIDS OF OXALATE OF LIME.   (Roberts.)

which polarize light very strongly.   (3) *Irregular discs*, which
have been likened to two kidneys with their concavities op-
posed, and sometimes so closely approximating as to appear
circular, the surfaces being finely striated, resembling the

coalescing spherules of a much rarer sediment—carbonate of calcium (Fig. 15). (4) *Well-defined* diamond-shaped crystals. The presence of organic matter, as mucus, may interfere with crystallization in the regular manner.   Again, the crystals may become entangled in the mucus, and thereby fail to fall to the bottom of the vessel, thus necessitating an examination of the mucus.   They have been seen impacted in the uriniferous tubules, and sometimes occur imbedded in, or deposited upon, tube-casts.   They are likely to be confounded only with abbreviated prisms of triple phosphate, carbonate of lime, or granules of uric acid.   Acetic acid dissolves the phosphates and carbonates (the latter with effervescence), and caustic potash the uric acid, while oxalate of lime resists the action of both of these reagents.

(To detect free oxalic acid in the urine, add to a filtered portion a few drops of acetic acid, and then a solution of a salt of lime, chloride for example.   Set the mixture aside for a few hours to crystallize, and examine the sediment.

Artificial Production.—Add to the urine, as above treated, a small lump of oxalic acid, and wait several hours for crystallization to take place.)

Phosphates.

Two-thirds or three-fourths of the phosphoric acid excreted by the kidneys (about 50 grains per diem) are combined with the alkaline bases, potash and soda, to form soluble phosphates.   These do not come under the notice of the practitioner as urinary sediments.   The remainder is combined with the earthy bases, lime and magnesia, to form salts, which, though soluble in *acid* urine, are speedily precipitated, whenever the urine becomes alkaline, as earthy phosphates.   These latter form calculi and urinary deposits.   There are three kinds, viz.:

1. Amorphous calcium phosphate ($Ca_3 PO_4)_2$, or bone-earth.
2. Crystalline calcium phosphate ($Ca H, PO_4$), or "stellar" phosphate.
3. Ammonio-magnesian phosphate ($Mg N_4 PO_4 + 6 aq$), or triple phosphate.

Amorphous Calcium Phosphate is most frequently found amorphous with the triple phosphate whenever the urine becomes alkaline. It is the normal deposit of the alkaline urine after a meal, and in persons whose urine has been rendered alkaline by remedies (carbonates, acetates, and citrates of the alkalies), and after the excessive use of subacid fruits. The turbidity caused by it exists in its greatest intensity at the moment of emission, and does not increase on cooling. It forms an amorphous whitish floculent deposit, indistinguishable by the naked eye from the cloud produced by mucus and epithelia. Having no affinity for the coloring matter of the urine, it is consequently paler than the supernatant liquid, differing in this respect from the amorphous *urates*. Heat increases this deposit. A drop of any acid causes it to immediately disappear. Its clinical significance is entirely involved in that of alkaline urine (see p. 17).

Crystalline Calcium Phosphate ("stellar" phosphate) is not a very common deposit, as compared with uric acid, the urates, oxalate of lime, or the amorphous and triple phosphates. It may exist in urine of a feebly acid reaction. Depressed acidity of the urine is an essential contingent to the formation of stellar phosphate crystals.

Microscopical Characters.—The crystals of crystalline or "stellar" phosphate of lime, present considerable variety of form. The prevailing appearance is that of crystalline rods or needles, either lying loose or grouped into stars, rosettes, or sheaf-like bundles, or into fans. Some of the crystals are club, wedge, or bottle-shaped, and abundantly marked with lines of secondary crystallization. It is also found in the shape of spherules or even dumb-bells. It is very frequently associated with oxalate of lime. These crystals are always

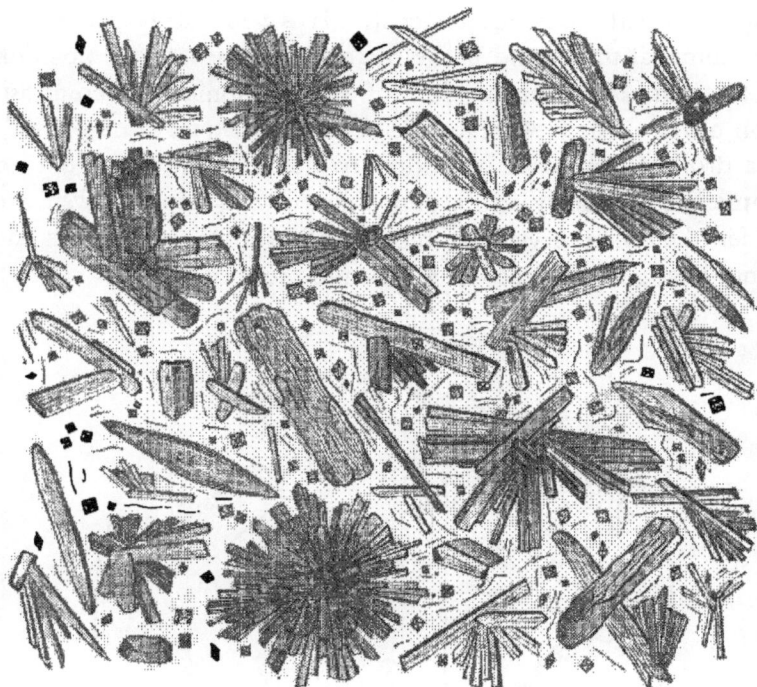

Fig. 16.—URINARY DEPOSIT.

Consisting of crystals of phosphate of lime, octahedra of oxalate of lime, with a little mucus. ×215. (Beale.)

colorless, and, being soluble in acids, are thus to be distinguished from one of the stellar forms of crystallization of uric acid.

The Ammonio-Magnesian, or triple, phosphate, occurs very frequently as a urinary deposit—sometimes alone, but much more commonly associated with the amorphous calcium phosphate. When unmixed with any other substance, this deposit has a snow-white appearance; and bright, glistening, glass-like crystals are seen studding the sides of the urine-glass, and forming a brilliant crystalline film on the top. Like the preceding deposit it may exist in urine that is feebly acid to test-paper. Heat does not affect it. It is necessarily present in ammoniacal urine, and in the majority of cases the deposition of this salt is only an incident of the ammoniacal decomposition of the urine. After a dose of salts (sulphate of magnesia), if a drop of ammonia be added to the urine it will cause an immediate turbidity and a deposition of the triple phosphate in fern-leaf like crystals, which, if allowed to stand for some time, change into those of a prismatic form.

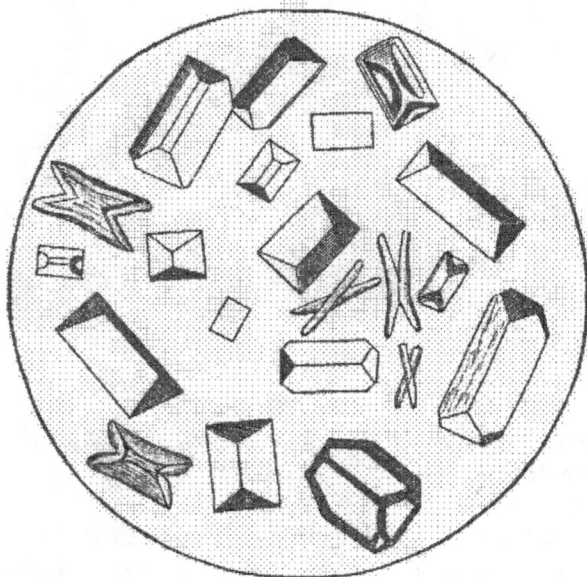

Fig. 17.—DIFFERENT FORMS OF TRIPLE (AMMONIO-MAGNESIAN) PHOSPHATE CRYSTALS. ×200. (Roberts.)

Microscopic Characters.—*The ammonio-magnesian* (or triple) *phosphate,* Fig. 17, has for the typical form of its crystals that

of a triangular prism with beveled ends.   By a planing-off of
the ridges and angles, and a hollowing out of the sides, a
great variety of subordinate forms are produced.   In highly
ammoniacal urine, the triple phosphate forms elegant di-elytral
crystals, which appears to arise from a hollowing of the sides
and a deep notching of the extremities.   Phosphates are, in
general, fine microscopic objects, but the appearances they
present greatly depend on whether they are slowly or rapidly
crystallized.   When the ammonio-magnesian phosphate crystallizes *rapidly*, as happens on the addition of ammonia to
freshly-passed urine, it assumes the form of fine feathers and
fronds ("fern-leaf"), either singly or in stellate groups (see
Fig. 18).   These feather forms show no play of colors with

Fig. 18.—CRYSTALS OF TRIPLE PHOSPHATE.
Formed by the addition of ammonia to the urine.   They are rapidly
developed.   If allowed to stand, they would gradually assume the prismatic form, as seen in Fig. 17.   ×215.   (Beale.)

polarized light, whereas the prismatic, and modifications of
the prismatic forms, are beautiful polariscopic objects, a selenite plate adding greatly to their beauty, for by its means a
fine-red field and deep-blue crystals may be obtained.   (Some-

times a scum forms on the surface of *acid* urine, composed of prismatic crystals much resembling those of triple phosphate in form, but larger in size. They are crystals of an organic substance, *creatinin*. They also form beautiful polariscopic objects.)

These three compounds are occasionally precipitated together in one deposit; much more frequently, however, the *first* and *third* will be found associated as a light-colored deposit, forming the ordinary sediment of ammoniacal urine.

Clinical Significance.—The presence of earthy phosphatic deposits in the urine signifies to the physician an alkaline or neutral condition of this fluid, the cause of which he should ascertain. If the urine contains a spontaneous deposit of earthy phosphates *the moment it is voided*, then the formation of vesical calculus is to be feared, especially if the deposit should contain crystals of *triple phosphate*. If the precipitate, under these circumstances, be *entirely amorphous*, the alkali causing it is *not* ammonia, and calculous formations, irritation and ulceration of the lining membrane of the bladder are not so much to be feared. It must not be forgotten that *amorphous calcium phosphate* is frequently present in the urine of those who are severe students, who lessen their hours of sleep, who exhaust their nervous systems, who use certain articles of diet, citrates, tartrates, carbonates (effervescing mineral waters). On the contrary, food or drink, rich in lime, as well as certain maladies, diseases of the spinal cord, and chronic vesical affections, are very apt to give rise to the crystalline forms of phosphates of lime. The urine of old people in general contains an abundance of phosphates. The appearance of a deposit of phosphates is no criterion of their absolute quantity, nor does a spontaneous deposit in fresh urine necessarily imply an excessive elimination of phosphoric acid. If the urine be abnormally acid, there may be an excessive elimination, and yet no deposit take place. Such a condition of affairs is liable to pass undetected. An *increase* of phosphates may be present in cases of phrenitis, injuries to the head; paralysis, especially when due to some disorder of the spinal cord; rickets, osteomalacia, extensive burns, in meningitis, cerebral tumors, tertiary syphilis, diabetes, and some forms of mania. A *diminution* of phosphates may be present in renal and intestinal affections, delirium tremens, gout, and severe pneumonia—their reappearance is a favorable sign.

Leucin and Tyrosin.

These two bodies belong to the same class; namely, transition products in the metamorphosis of nitrogenized substances. They are *very rare* deposits in the urine, being found only in urine which is loaded with biliary coloring matters, since they generally attend only grave destructive diseases of the liver, especially acute yellow atrophy and phosphorus poisoning. Harley says of leucin that all the cases in which it has yet been detected have terminated fatally; and of tyrosin, "its presence in the urine is an almost certain sign of a rapidly approaching fatal termination." Urea is almost always diminished in urine containing these substances.

Microscopical Characters (Fig. 19).—*Leucin* presents itself as more or less yellowish-tinged, highly-refracting spheres or globules, which may, at first sight, be mistaken for *oil-drops*. They do not, however, refract light quite so strongly, not having so wide a dark border; and by suitable illumination, many of them will be found marked by radiating and concentric striæ. These spherules exhibit a peculiar disposition to aggregate, appearing partly to merge where two edges come together. They might possibly be mistaken for crystals of *carbonate of lime*, but the latter sink, whereas the greater number of globules of leucin float in water.

Tyrosin crystallizes in the shape of very fine, long, delicate, silky, acicular prisms, arranged in beautiful bundles, tufts, or sheaf-like collections, often crossing each other, forming a

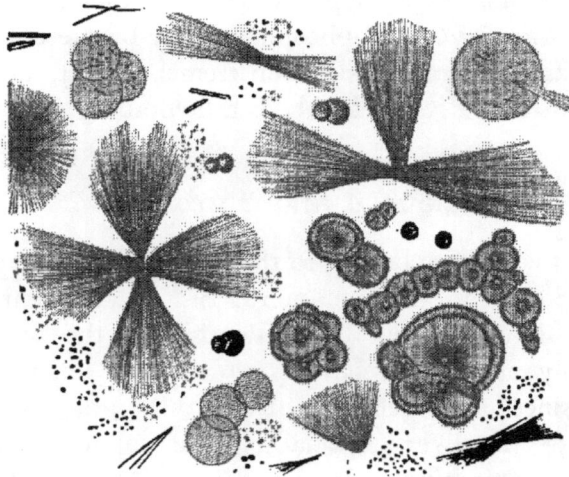

Fig. 19.—LEUCIN SPHERES AND TYROSIN NEEDLES.   (Tyson).

cross of four bush-like arms, and intersecting at their constricted central portions; or they may form spiculated balls, not unlike a rolled-up hedgehog with the bristles sticking out in all directions.   The crystals found in urine are usually of a deep yellow color from absorbed bile pigment.

Cystin.

Cystin (*cystic oxide*) ($C_3 H_7 NSO_2$)

Cystin is never seen in normal, and but rarely in diseased, urine, while, as compared with other urinary sediments, it is one of the most rare.   The production of cystin has as yet not been fully elucidated, but it is supposed to be in some way or other connected with the function of the liver. Urine depositing cystin is usually feebly acid, of a yellowish-green color, oily appearance, and a peculiar sweetbriar odor. When voided it is turbid, but on standing, a copious light sediment subsides, having much the appearance of fawn-colored urates, but which is unchanged by heat. Cystic urine is very liable to spontaneous decomposition, in the course of which it evolves ammonia and sulphuretted hydrogen (derived from the 26 per cent. of sulphur which it contains), and blackens white glass vessels. Cystin may exist alone, but it is more commonly observed in urine with ammonio-magnesian phosphates, mucus, and epithelia. A greasy-looking pellicle, consisting of the former salt and cystin, soon forms on cystic urine. When held in solution in neutral or alkaline urine, a deposit may not occur until *acetic acid*, which precipitates it, has been added.

Microscopical Appearances (Fig. 20).—A spontaneous deposit of cystin in the urine is usually composed of regular colorless hexagonal tablets of different sizes, united by their flat surfaces, and overlapping one another. They may have an iridescent, mother-of-pearl lustre; their surfaces are often chased by lines of secondary crystallization; they also form thick rosettes of great brilliancy, with sharply crenate margins, darker in the centre than at the circumference. They color polarized light. Being, however, a dimorphous body, cystin may occur in square prisms, either singly or in stars. These prisms refract light strongly, and the facettes which lie slantingly out of the direct line of vision appear perfectly black, contrasting with the brilliant lustrous white of the planes through which the light passes vertically. Cystin is instantly dissolved, without decomposition, by ammonia, and if the solution be exposed in a watch-glass to evaporation in the air, beautiful six-sided crystals appear as the volatile alkali dissipates.   This is the characteristic reaction of cystin, and leads to its easy identification.   From the rarely-occurring, hexagonal-looking plates of uric acid, cystin crystals may be distinguished by their lesser size, their absence of color, and their solubility in a mineral acid, from which they

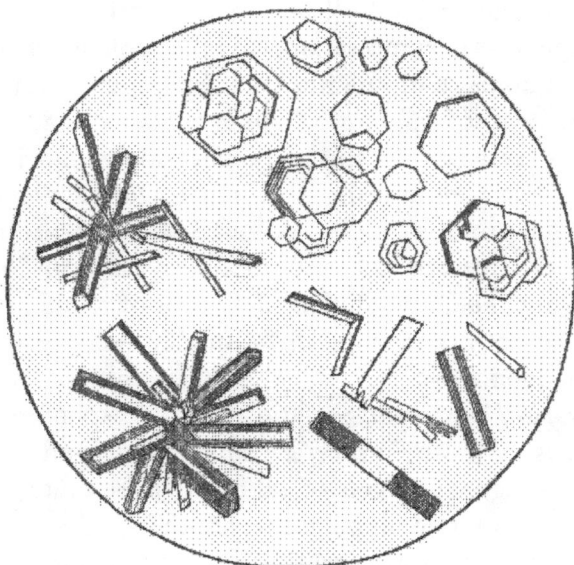

Fig. 20.—CYSTIN, HEXAGONAL TABLETS AND PRISMS.   (Roberts.)

are redeposited on the neutralization of the acid by ammonium carbonate.   The ammonio-magnesian phosphate often accompanies the crystals of cystin.

Clinical Significance.—The chief, if not the whole, clinical significance of cystinuria lies in the danger of the formation of cystic calculus or gravel.   Cystinuria has been known to exist for years without any other deviation from health than that caused by the physical irritation of the concretions when these form.   It is believed by some authors to be hereditary.

Xanthin.

Xanthin $C_{10} H_4 N_4 O_4$.

Xanthin is rarely met with in diseases as a urinary sediment, and still more rarely as a calculus. It is supposed to be one of the intermediate products of the metamorphosis of protein substances. Neubauer thinks that on an average there is about 1 gramme (15·5 grains) in 600 pounds of human urine. Pure xanthin is white, but as met with in the form of calculus or deposit it is of a yellow or dark cinnamon-red color, due to its taking up a quantity of the coloring matter (urohæmatin) of the urine, just as uric acid under similar circumstances does.

Microscopical Characters.—Small oblong plates resembling

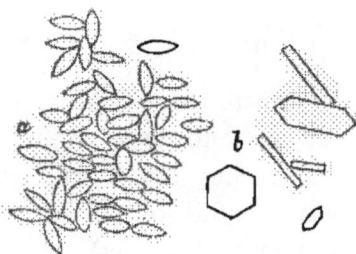

Fig. 21.—XANTHIN FROM HUMAN URINE.

(*a*) Spontaneous deposit. (*b*) Dissolved in hydrochloric acid and re crystallized. (Harley.)

uric acid. They are soluble in ammonia, caustic potash, strong mineral acids (with the aid of heat), and insoluble in cold water, ether, and alcohol.

CHAPTER IV.

ORGANIZED URINARY DEPOSITS; MUCUS; EPITHELIUM; RENAL
TUBE CASTS; BLOOD; PUS; SPERMATOZOA; FUNGI.

In the majority of cases, but not invariably, when urine is
voided turbid, or shows a sediment before cooling, it will be
found to arise from one or more of the *organized* urinary deposits, as mucus, epithelium, renal tube-casts, blood, pus,
spermatozoa, or fungi.

Mucus.

Mucus is a constant constituent of urine, and if perfectly
normal urine be allowed to stand for an hour or two, a *very*
light, delicate, semi-translucent cloud will be seen floating
rather in the lower fourth or fifth of the fluid than actually
upon the bottom of the vessel (Fig. 22). If now, by means
of a pipette, a small portion of this suspended flocculent cloud
be placed upon a glass slide, and then examined with a power
of from 200 to 500 diameters, a few rounded or oval cells,
leucocytes (mucus corpuscles), about 1-2500 of an inch across,
together with a small number of epithelial cells, or remnants
of the same, detached from the various surfaces of the urinary-
renal tract over which the urine has passed, will be seen
entangled in the *invisible* net of mucus. As a rule, of mucus
having the usual glairy character, there is no visible trace in
healthy urine, since *mucin*, being so transparent and similar
to urine in its refractive index, could not be recognized by its

own properties. Among other substances, acetic acid precipitates the mucin in the shape of pale, delicate, fibrillated,
or punctated bands, or threads, which are sometimes tortuous,

Fig. 22.—THE SO-CALLED "MUCUS CLOUD."

Consisting of mucus corpuscles and epithelial cell. Exaggerated for
the purpose of demonstration. (Harley.)

and frequently anastomosed. If a little iodine or iodide of
potassium be added to such acetic acid, the mucin is both precipitated and colored, while the epithelial cells and leucocytes
are rendered more distinct. These coagula (or mucus-casts,
as they have been called) may sometimes be found in urine to
which no acid has been added; found studded, perhaps, with
granules of the urates, and, under these circumstances, may
be mistaken for granular tube-casts. They are, however, generally very much narrower than the latter, have badly-defined
margins, and the urate granules are dissolved by warmth or

hydrochloric acid.   Should any part of the urinary renal tract
become irritated, the mucus-cloud will become more opaque
and greatly increased in bulk, and an abundance of leucocytes will appear, accompanied by numerous epithelial cells,
indicative of the part from which they came.   Should the
irritation pass on to inflammation (the causes producing pus
and mucus are but differences of degree), pus will appear in
the urine.   So long, however, as urine containing mucus is
non-albuminous, pus may be said to be absent, *since pus contains albumen while mucus does not.*   Mucus appears to be an
important factor in the fermentative processes of the urine.
(*See Pus.*)

Epithelium.

Any part of the genito-urinary tract may shed its epithelium
into the urine, so as to form a sediment.   The urine of the
two sexes differs notably in the character and quantity of the
epithelial cells found therein.   This arises from the anatomical
differences in the lower genito-urinary passages; and advantage may sometimes be taken of this circumstance to distinguish the sex of the individual whose urine is under examination.   It is not very often that one is enabled to locate
the source of epithelium coming from beyond the bladder and
vagina, partly because of the comparatively slight differences
in the epithelial cells from certain locations, and partly
because maceration in the urine causes the cells to distend,
change their shape, and thus render feebly distinctive points
even less marked.   It may here be observed, in passing, that
in acid urine epithelial cells remain for a considerable length
of time, but in alkaline urine they are gradually destroyed,
becoming at first more swollen and transparent.   The epithelium from the *urethra* is of the *scaly* form, in the vicinity of
the meatus, *columnar* higher up, and around the prostate

the cells are caudate, spindle-shaped and irregular. The epithelium from the bladder varies according to the part of the organ from which it is derived; from the *fundus* it is *columnar* mixed with large oval cells, which are considerably larger than round cells from any other source; from the

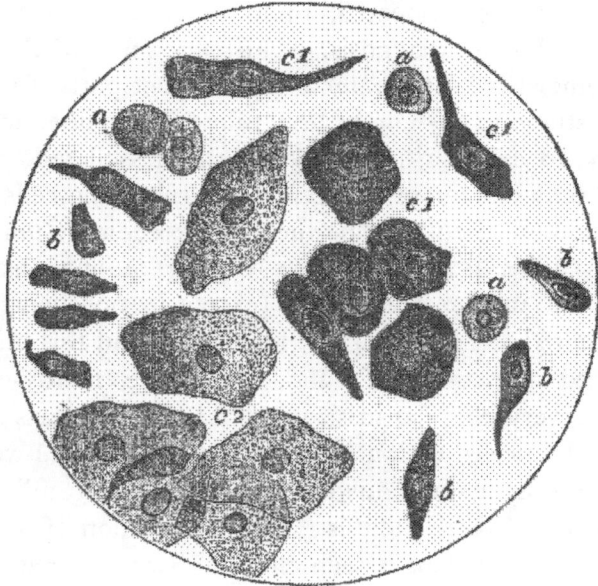

Fig. 23.

a, round epithelium from bladder; *b*, columnar epithelium from ureter and urethra; *c*1, columnar and squamous epithelium from deeper layers of epithelium of bladder; *c*2, squamous epithelium from superficial layers of epithelium of vagina. (Tyson.)

trigone, the cells are large, slightly flattened, with a very distinct nucleus and nucleolus, but not so large or as flat as those from the vagina; from the *ureters* the cells are of the *columnar* and spindle-shaped form, with a large distinct nucleus. The epithelium from the *pelvis of the kidney* is composed in part of the *tessellated*, and in part of the *columnar* variety, like that of the ureter. (*c* 1) Some of these cells so closely resemble cancer-cells as to render the recognition of cancer-cells (as such) in the urine a matter of great uncertainty. All of the above-mentioned epithelia find their way into the urine in cases of renal calculus, pyelitis, and vesical calculus; and where pyelitis is suspected, their appearance in the urine greatly fortifies the diagnosis. The epithelium of the kidney differs somewhat in its characters in different parts of the tube. In the straight portion of the tube the epithelium has a tendency to flatness approaching the tessellated variety; while in the convoluted portions of the tubes, the cells consist of a round or slightly oval nucleus, exceedingly uniform in size and shape, having a delicate regular outline resembling closely both in size and aspect (except in not being bicon-cave) the red corpuscle of the blood; around this nucleus is aggregated a quantity of solid, yet friable, faintly-granular substance, very irregular in size and shape, sometimes forming a spheroidal mass, with a more or less distinct cell-wall, and sometimes only a small quantity of it remaining adherent to the nucleus. Where there is rapid proliferation of the renal epithelium, the nucleus is seen cleft into two or three nucleoli, and the cell takes on the appearance of a pus-corpuscle. But, as a rule, they are distinguished from pus-corpuscles by their larger size, their *single* nucleus, which is distinct without the use of reagents, while the multiple nucleus of the pus-cell re-quires the use of acetic acid to bring it out.

Epithelium.—This table is only an approximate guide for the microscopic diagnosis of epithelia; intermediate forms and dimensions exist that escape any classification:

1. Round or oval epithelial cells, a little swollen.	Very large (o mm. ·016 to ·033 with a single large nucleus, ordinarily o mm. ·011).	*Urethra.*
		Bladder (trigone).
	With 2 nuclei, or one nucleus, and nuclei form granulations.	*Pelvis of the kidney.*
		Ureters (superficial layer).
	Much smaller (about o mm. ·011 to o mm. ·015).	*Kidney (rarely in an isolated state).*
		Bladder (near the neck).
2. Epithelial cells lamellar, very thin, polygonal.	A large nucleus, and often two nuclei and nucleiform granules.	*Pelvis of the kidney and Ureters (superficial layer.*
	Very large lamellæ (o mm. ·022 to o mm. ·045), but with very small nucleus (o ·006).	Vagina and external genital parts.
		Urethra, near the meatus.
3. Cylindrical, Columnar, spindle-shaped, epithelial cells, with tail longer or shorter, and more or less bent.		*Bladder. Ureters. Pelvis of the kidneys.*

In females, epithelial sediments are both common and abundant. From the short urethra the urine receives little or nothing; but the vaginal mucus membrane, invested as it is throughout with a lining of pavement epithelium, the elements of which are detached in great quantity with facility, gives rise to an abundant, amorphous-looking, light, cloudy deposit, which, under the microscope, is seen to be composed of large, flat, mono-nucleated cells, often thicker at the middle, irregularly polygonal in outline, and often folded over themselves either completely or partially; occurring either discrete, or united by their borders into patches of rude mosaic (*c2*, Fig. 23.) A deposit of this kind is found almost invariably and only in the urine of females, being very abundant in vaginal leucorrhœa, and even where there is no appreciable disorder of the parts, as in the case of young female children of a strumous diathesis.

Since, then, there is no positively exact way of differentia-
ting the precise sources of these round epithelial cells, other
things must be taken into consideration; for example, the
fact of their being found in *albuminous* urine, together with
casts, makes it probable that they came from the uriniferous
tubules; if there be symptoms of impacted calculus, from the
pelvis of the kidney; otherwise from the urethra, or Cowper's,
or the prostate gland.

An increased amount of epithelium from the urethra indicates catarrhal or specific inflammation of the walls of this
canal; from the bladder, more or less catarrhal inflammation
of the lining membrane of this viscus; from the vagina leucorrhœa, or specific inflammation; from the ureters or kidneys, a more or less serious affection of these organs, according to the amount of cells present, in connection with
albumen, blood, renal casts, pus, and local and constitutional
symptoms.

Renal Tube-Casts.

Collection and Examination.—In congestion, and in inflammation of the kidney (Bright's disease), there are formed, in
the uriniferous tubules, cylinders or moulds, which are discharged with the urine, and form the deposit known as
" casts." When the urine contains casts in great abundance,
they can scarcely be overlooked, if, after the urine has been
allowed to settle for a few hours in a tall *cylindrical* glass, the
whole of the supernatant liquid is poured off, and then one
of the last drops which flows from the lips of the vessel be
put under the microscope, on a slide furnished with a *shallow*
(gum-dammar or cement) *cell*, and examined with a comparatively low power, which, say with a ½ in. or 4-10 in.
objective (100 to 250 diameters), will include at once a

considerable quantity of urine. Once found, they can subsequently be studied with a much higher power (1·5 in. to 1·8 in. objective; 400 to 800 diameters). If there are but few casts present in the urine, there may be no deposit appreciable to the naked eye. In this case, collect the whole of the urine of the twenty-four hours, put it in a tall cylindrical glass, let it stand from 12 to 24 hours to settle (the longer, the more albuminous the urine is); then pour all the urine away, except the four or five ounces which have collected at the bottom. Allow this residue to settle again for a few hours, when, if they be present, casts will be found at the bottom of the glass, whence they may be taken by means of a pipette, or one of the last drops poured off may be used, and examined as above described. The search must not be relinquished until several such slides have been carefully and patiently examined, nor must the case be pronounced upon until several examinations of the urine have been made at intervals of several days.

Classification.—Casts may be conveniently divided, according to their appearance under the microscope, into *three* kinds : *epithelial* casts, *granular* casts, and *hyaline*, or waxy, casts.

Origin, or Mode of Formation.—There are three views as to the mode of formation of casts : *first*, that they are a secretion from the epithelial lining of the tubules; *secondly*, that they result from a desquamation and degeneration of the epithelial cells; and *thirdly*, that they are formed by the escape, by capillary rupture or otherwise, of blood, or a coagulable constituent of the blood into the tubules, where it solidifies, entangling, as it does so, whatever it may have surrounded in its liquid state, such as blood corpuscles, leucocytes, epithelium, uric acid, urates, and especially oxalate of lime crystals; that it then becomes moulded to the shape of the tube into which it has been extravasated; subsequently it contracts and slips out of the tubule into the pelvis of the kidney, whence it is carried through the ureters to the bladder, and voided with the urine. The first of these views is generally discarded; the second and third are viewed with favor as accounting for epithelial and granular casts, while it is probable that the great majority of hyaline and waxy casts

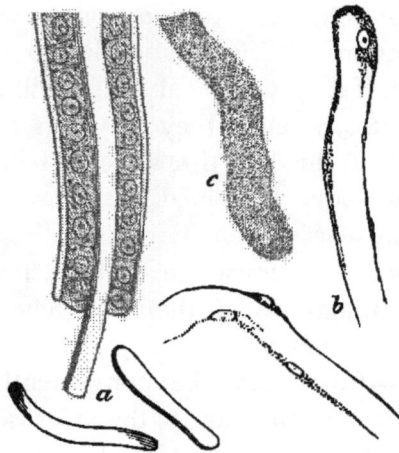

Fig. 24.

a, b, hyaline and (c) granular casts, illustrating the formation of the former at a. (Tyson.)

are formed in the last way described. Probably most of the casts found in the urine are formed in the straight uriniferous tubules (Fig. 24).

Epithelial Casts (Fig. 25, *a a*).—These are casts consisting of a mass of epithelial cells derived from the tubules of the kidney. They are usually wide, never very narrow. The cells may undergo molecular change and give rise to

The Granular Cast (Fig. 25, *b b*).—This is a solid cylinder, having a dark, almost black, coarsely-granular aspect;

or else dotted here and there with a few dark points in its substance. Granular casts may occur in all forms of albuminuria, but the cause of their appearance is always *active*

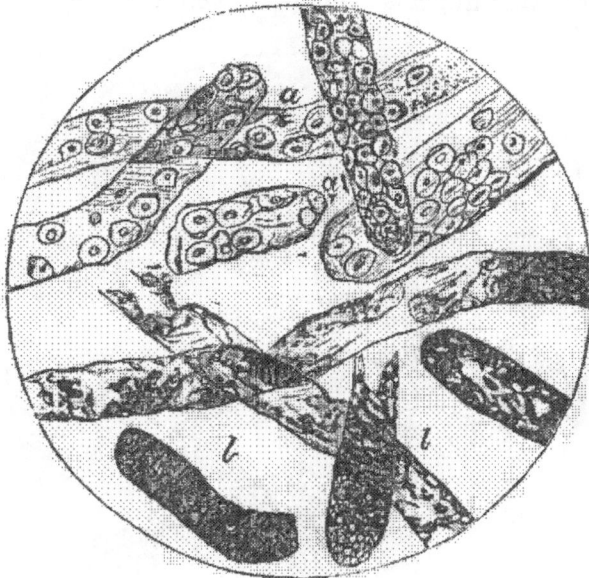

Fig. 25.

a a, "epithelial" casts; *b b*, "opaque granular," from a case of acute Bright's disease. (Roberts.)

congestion (Harley). The "fatty or oil-cast" is a variety of the granular produced by the coalescence of the olein granules into globules (Fig. 26, *a a, c*). Granular casts may appear as transparent cylinders studded with tolerable uniformity with minute oil-globules. When blood corpuscles exist on these casts, they are called "blood-casts" (Fig. 26, *l l*), and appear as perfect cylinders, composed of delicate circles placed in opposition; more usually as a fibrinous cylinder irregularly studded with blood corpuscles, some perfect and some shrunken and contorted, and, finally, the cast may appear to be composed of nothing but blood corpuscles, crushed or compressed into a cylindrical mould. When the material of granular casts is derived from broken-down blood corpuscles, the cast appears yellow or yellowish-red. Granular

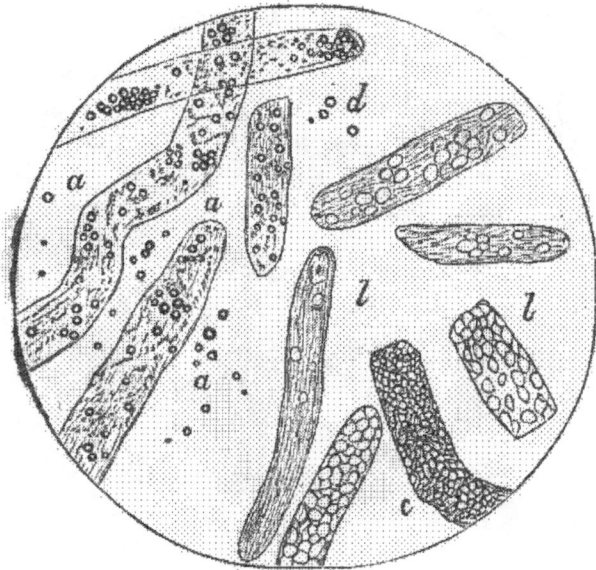

Fig. 26.

a a, fatty casts; *l l*, blood casts; *d d*, free fatty molecules. (Roberts.)

casts may be large or small, full, or only partially filled with granular matter; the size of the cast depends on the part of the tube in which it was formed, the amount of granular matter it contains in the condition of the kidney.

The Hyaline, Structureless, or Waxy Casts (Fig. 27).—These are usually very clear cylinders, whose outlines are often so very delicate and indistinct that, before they can be detected, they must be artificially tinted with iodine or magenta; or the light from the mirror illuminating the field of view must be modified, by shading with the hand, or by manipulating the mirror itself, the field of view being just outside the exact focus. This method is peculiarly applicable to the finding of the so-called "waxy" casts, which, owing to their greater refractive power, will so concentrate the light as to become much more visible. Completely hyaline homogeneous casts do occur; but, generally, what are called hya-

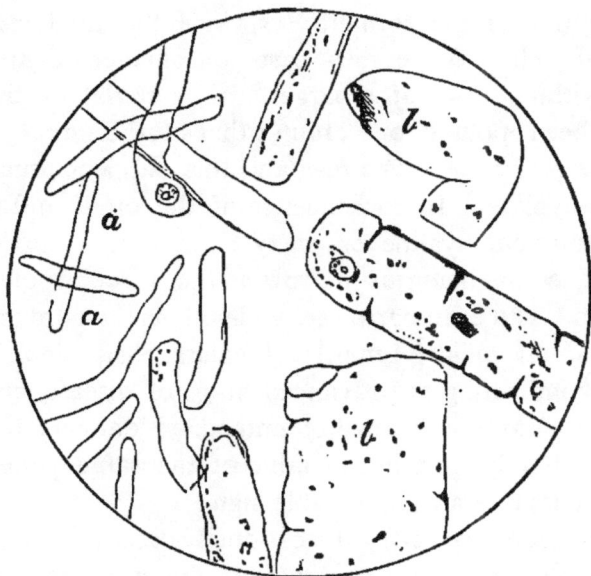

Fig. 27.—WAXY CASTS.

a a, urine of man with chronic Bright's disease; *b b*, chronic Bright's disease (large white kidney); *c*, chronic Bright's disease (contracted kidney with fatty degeneration). (Roberts.)

line casts shows indistinct markings on the surface, or a faint molecular composition by a few granules here and there, or one or two glistening oil-drops. A hyaline cast of more solid aspect, resembling molten wax, is termed a "waxy cast." Hyaline casts present extreme differences in diameter, according to the manner and place of their formation. If, when the transudation occurred into the tubule, the epithelium was so firmly attached to, the basement membrane as to remain behind when the cast passed out, then we will have a *narrow* hyaline cast, which may sometimes be of considerable length (Fig. 24, *a*). If, on the other hand, the tubule was entirely denuded of its epithelium, then a wider one; the diameter of the latter being equal to that of the former, *plus* twice the thickness of an epithelial cell. Sometimes casts of smaller diameter are found within those of a larger, the material of the latter having been poured out around that of the former, *after it had undergone some contraction*, and this happens usually with waxy or hyaline. In consequence of the mode and place of their formation, hyaline or waxy casts vary considerably in diameter, some being as narrow as the 1-1000th of an inch (.025 mm.), and even narrower, while others are as much as the 1-50th of an inch (·5 mm.) Hyaline casts alone, though generally indicating an advanced stage of disease, are by no means invariably indicative of immediate danger; but associated with oily casts in the urine of the same patient they always constitute an unfavorable sign.

Mucus Casts (Fig. 28).— Occasionally casts are found which are apparently pure *mucus-moulds* of the uriniferous tubules, and unless they are covered by accidental elements (as granular urates or phosphate of lime), they are smooth, hyaline, faintly-fibrillated cylinders, especially characterized by their *great length* (which is often enormous), in the course of which they divide and subdivide, diminishing in diameter as the division proceeds, showing conclusively a renal origin. They are usually associated with *pus*, since they are particularly apt to occur where there is vesical irritation extending through the ureters to the kidney, and there is no more albumen than can be accounted for by the presence of the pus.   Dr. Beale says they are not unfrequently seen in urine of a very high specific gravity (1030 and higher) containing an excess of urea and urates.

Fig. 28.—MUCUS CASTS.

From the straight portion of the uriniferous tubules, showing the manner in which the large renal tubes divide and subdivide as they pass towards the base of the pyramids. ×75. (Beale.)

Casts of the Seminal Tubules are sometimes found in the urine, but their origin may be inferred from the presence of spermatozoids in them.

Diagnosis.—In diagnosticating veritable tube-casts from other objects simulating them in appearance, the following points will aid in their determination : (*a*) the discovery of epithelial cells from the uriniferous tubules, or of red-blood corpuscles and leucocytes, *imbedded in or firmly adherent to the cast ;* care must be taken not to mistake cells that simply overlie the cast and are not imbedded in it.   To accomplish this, cause a to-and-fro current in the liquid by gently tapping with a mounted needle upon the cover, and thus separate the two objects, if the union between them is not intimate. Again, many casts occur without any epithelial cells upon them at all (as in the last stages of albuminuria, when some of the tubules have been completely stripped of their lining epithelium); but if the search be sufficiently prolonged it is highly probabe that one or two tube-casts, exhibiting well-defined epithelium, will be found. (*b*) The size of a suspected cast will often give assistance. Casts ordinarily vary from 1-1400 to 1-500 of an inch across, and the cases where diagnosis is difficult are those where the casts do not exceed the 1-1000th of an inch in diameter; because where the disease has so far advanced as to strip many of the tubules of their epithelial lining, and so allow large and medium casts to be formed, a well-marked albuminous precipitate will be found on testing the urine with heat and nitric acid, and the history or condition of the patient (without being acquainted with which no physician should give a *positive* opinion) is usually such as to throw sufficient light on the subject and prevent erroneous conclusions. (*c*) Though not an invariable character, veritable tube-casts have *round or club-shaped* extremities.

Clinical Significance.—The presence of casts in the urine is a sure sign of renal disease, but not, however, necessarily of a *permanent* disease. They are present in many acute diseases, accompanied by albumen in the urine. If found for several weeks together, unaccompanied by pyrexia, the existence of some serious disorder in the kidneys is to be inferred. They are present in all cases of renal congestion, and all forms of Bright's disease. Some assistance may be derived from the appearance of the casts in forming a judgment of the acute or chronic character, or a diagnosis and prognosis of a disease. A deposit may, and generally does, contain a mixture of two or more varieties of casts and cells; and hence, conclusions as to their pathological meaning must be deduced from the *prevailing* types rather than from the absence or presence of one or two of a given character.   Bearing in mind these precautions, and having regard to the previous history of the case, the following conclusions are generally warranted: (*a*) *Epithelial casts*, of medium size, especially when they have undergone little or no granular change, together with a large quantity of epithelial cells, and also *blood-casts*, indicate a disease of recent origin—congestion, acute Bright's disease.   (*b*) *Fatty* or intensely *granular casts*, with scanty epithelium, the cells withered and contracted, or containing globules of oil—chronic Bright's disease, fatty degeneration.   (*c*) Transparent, large *waxy casts* (over 1-600th of an inch in diameter)—chronic disease in a very advanced stage.

Blood in the Urine—Hæmaturia.

Occurrence and Origin.—Blood is not at all infrequently found in the urine, and it may be derived from any part of the urinary renal tract.   Unless the quantity be very small, it imparts a color to the secretion when mixed therewith.   If the blood is derived from the kidneys it is equally diffused through the urine, to which it imparts, if the urine be acid, a peculiar, quite diagnostic, "smoky" hue; if the urine be alkaline, or the blood in great quantity, a pinkish, bright red, or vermilion color.   In the former case, after standing awhile, a chocolate-colored grumous deposit subsides.   If the source of the hæmorrhage be in any part of the urinary tract *below* the kidneys—ureters, bladder, and urethra—the color will be that of the blood itself, and frequently distinct clots will be found in the deposit.   If the urine be acid, and of moderate density (1020 to 1025), the blood corpuscles, like epithelial cells, will remain visible and preserve their form for several days.   But in an ammoniacal urine, of low specific gravity (and bloody urine of low density is prone to speedily become alkaline and decompose), they soon disappear, having been dissolved by the alkali. The hæmato-crystalline and hæmatin are then dissolved out into the urine, and may be tested for chemically. *Urine containing blood in sufficient quantity to be recognized by the naked eye, is always albuminous.*

Microscopical Characters.—If the urine be examined under the microscope, before the blood corpuscles have perished, the presence of blood can easily be made out. Blood corpus-

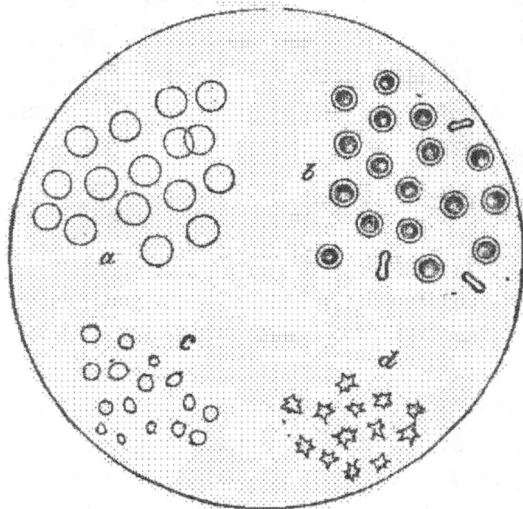

Fig. 29.—BLOOD-CORPUSCLES IN URINE.

a, slightly distended by imbibition; *b,* showing their bi-concave contour; *c,* shrivelled; *d,* serrated. (Roberts.)

cles in the urine do not run into rouleaus, as they do when drawn directly from a blood-vessel, but they are separate. Moreover, long-continued maceration tends to interfere with their well-known optical properties (due to their bi-concave shape) of reversing light and shadow, when focussed alternately from their centre and periphery. This interference depends very much upon the density and reaction of the urine. In acid urine approaching a specific gravity of 1028 (that of blood-serum), the red and white blood corpuscles will often be found, if examined soon after being passed, presenting a perfectly normal appearance. If the urine be of a low specificic gravity (1010 or less), the red corpuscles will swell up, from endosmosis, becoming biconvex instead of biconcave (no longer now reversing light and shadow), then spherical, perhaps rupturing, and finally so colorless (from the exosmosis of their hæmato-crystalline) as to become almost invisible to the unpracticed eye, if not aided by a very superior or very powerful lens, or unless their transparent cell-wall has been colored by a little iodine or magenta, or shrivelled by adding a drop or two of a concentrated solution of sodium sulphate to them. These circumstances nearly double their ordinary size, varying from 1-2000th to 1-1400th of an inch in diameter, and showing a distinct cell-wall, and single, double, or multiple nuclei, around which, under a high power, may be seen numerous rapidly-revolving molecules. These changes of shape and size as the density of the circumambient fluid is altered, will generally enable us to distinguish red blood corpuscles from the spores of certain fungi, which resemble them under low powers. Red corpuscles are non-nucleated, while, with a good glass, a nucleus or vacuole may be always detected in the sporules, which are, besides, oval or elongated. If these fail, set the urine aside for a day or two, when the spores of such will manifest themselves by germination, or budding. From the nuclei of renal epithelium, red blood corpuscles may be distinguished by their feeble refractive power, and their being less strongly colored by magenta.

Clinical Significance.—Hæmaturia may arise from a great variety of causes, which may be classified as follows:

1. *Local Lesions*—External injury, violent exercise, calculous concretions, ulcers, abscesses, cancer (or fungus growths), tubercle, parasites (as Bilharzia hæmatobia), active or passive congestion, Bright's disease.

2. *Symptomatic*—In purpura, scurvy, eruptive and continued fevers, intermittent fever, cholera, etc., mental emotion.

3. *Supplementary or Vicarious.*—To hæmorrhoids, asthma, menstruation. In females the urine is generally bloody during the menstrual flow; it may also become so at any time if there be uterine and vaginal hæmorrhage. In testing for *albumen*, under these circumstances, it may be necessary to obtain the urine directly from the bladder by means of a catheter. If the amount of blood in the urine be small, the chances are in favor of its coming from the kidneys, and search must, therefore, be made for the accompanying tube-casts. If the amount be large, it probably comes from the pelvis of the kidney, ureter, or bladder; if from the pelvis and ureter, there will also be pus, and probably gravel, in the urine, with lumbar pains shooting down the thigh and testicles. If none of these indications be present, the hæmorrhage is probably vesical in origin. The chief danger of hæmaturia is the formation of clots in the passages, and consequent ischuria. Small clots may sometimes form the nucleus of a calculus.

Pus in the Urine—Pyuria.

Occurrence and Recognition in the Urine.—Pus is often present in the urine, causing it to be milky and turbid when voided, and afterwards falling as a dense and opaque yellowish white sediment to the bottom of the urine-glass. The turbidity of the purulent urine is not affected by heat. The urine readily becomes alkaline, and rapidly decomposes after being passed. So long as the urine remains acid, the deposit is loose and the corpuscles discrete; but when the urine becomes alkaline, as it often does, from ammoniacal decomposition, the pus corpuscles are destroyed, as to their morphological identity, and the pus coheres into a grayish viscid tenacious mass, which can be drawn out into long tough strings. This ropy appearance is diagnostic of pus. In this condition the pus-corpuscles have been destroyed. If a fluid drachm of the sediment, the supernatant liquid having been poured off, is mixed with an equal bulk of liquor potassæ in a test-tube, it will become transparent, and so gelatinous and tenacious that the vessel may be inverted without the evacuation of its contents. Urine containing mucus becomes more fluid and limpid by the addition of caustic alkali. Purulent urine contains, necessarily, more or less of *albumen*, according to the amount of pus present, from a trace too slight to be detected by ordinary reagents, to considerable impregnation. Filter before testing for the albumen.

Microscopical Characters (Fig. 30).—Under the microscope (300 or 400 powers) the deposit is found to be composed of spherical cellular bodies (pus corpuscles), about one-third larger than a red blood corpuscle; that is, 1-3000th to 1-2500th of an inch (·008 to ·010 mm.) in diameter, having granular contents and nuclei, varying from 1 to 4 in number. As stated, in speaking of blood in the urine, the pus corpuscles (leucocytes) vary in size with the density of the fluid in which they occur. The denser the urine the smaller and more crumpled becomes the pus-corpuscle (1-3000); whereas, the addition of water expands and clears it (1-1400), sometimes bringing into view the nucleus. A drop of dilute acetic acid (20 per cent.) placed upon the slide at the margin of the cover, and allowed to flow beneath it, brings out the neuclei very soon and with great distinctness, while with a high power (or a superior objective of low power, ¼th inch) the delicate cell-wall, of extreme tenuity, may be seen surrounding the nuclei, until after a time it bursts and disappears.

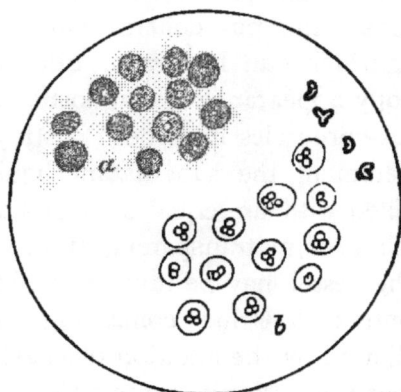

Fig. 30.—PUS CORPUSCLES.

a, without reagents; b, after the addition of acetic acid. (Roberts.)

Carmine solution, added to the urine, colors the nuclei of leucocytes.   If the acid be added in excess, the cell-wall and contents disappear altogether, and the nuclei float free in the .fluid.

Clinical Significance.—This depends on its source and quantity.   Suppuration may occur at, or abscesses burst into any part of the urinary renal tract, and cause pus to appear in the urine.   Purulent urine occurs in the following diseases: In men, *gonorrhœa* or gleet is the commonest cause of purulent urine; it is always small in quantity, and does not affect the general properties of the urine.   In women, leucorrhœa is the most common cause, and is manifested by the co-existence of an abundance of pavement epithelium from the vagina.   Pus from the bladder, *cystitis*, is of more serious significance. Origin determined by the local symptoms. The alkalinity of the urine in these cases may cause the gelatinization of the pus into a viscid mass, even while yet in the bladder, and thus give rise to excessively painful micturition, or the change may occur just after the urine has been voided. Triple phosphates occur in such urine. *Pyelitis*, suppuration in the pelvis of the kidney, is generally indicated by transitional epithelium, acid reaction of the urine, absence of signs pointing to the bladder or urethra, and lumbar pains. The bursting of an abscess into the urinary passages is manifested by a *sudden* irruption of a large quantity of pus into the urine.

Spermatozoa in the Urine.

These little bodies are frequently present in the urine, where, with the seminal fluid, they form a glairy white deposit, but generally there is little or nothing in the appearance of the urine to indicate their presence. They can be detected only by means of the microscope; using a low power (250 diameters) at first, to rapidly pass the suspected deposit under review, should only a few of them be present. Subsequently, they require a magnifying power of from 400 to 600 diameters (one-fifth objective with a No. 2 eye-piece) to show them well. They are recognized by a characteristic oval or pear-shaped anterior enlargement (the head or body) 1-3000th of an inch in length, and a long, extremely delicate, finely-tapering appendage (the tail) averaging 1-25th of an inch in length; all these parts appearing completely homogeneous, possessing a peculiar bluish tint, and a kind of fatty lustre, which renders it, like the spores of the yeast-plant and some hyaline tube-casts, most easily detected when viewed just outside of or beyond the exact focus. In semen freshly-passed they exhibit active eel-like movements, strongly suggestive of volition. Water or acid fluids completely arrest the movements of the spermatozoids, and hence, when seen *in urine,* they are always *motionless.* In healthy alkaline vaginal mucus their peculiar undulatory vibrations are plainly visible. They possess the power of resisting disintegration to such a

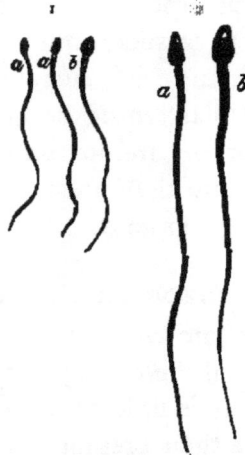

Fig. 31.—HUMAN SPERMATOZOIDS.

1 ✕ 350. 2 ✕ 800 diameters. *a,* viewed from the side; *b,* from the front.

remarkable extent that they have been found in putrefied urine at the termination of three months. Mineral acids and caustic alkalies attack them only when heated; acetic acid only renders them more conspicuous, while, after drying and subsequent softening in water, they preserve a recognizable and perfectly characteristic form, which may be of great importance in medico-legal examinations. Solutions of carmine, iodine water, etc., which color leucocytes and epithelia, have no action on spermatozoids. Ammonia, however, rapidly acts upon them, even when cold. In cases where they are mixed with certain deposits, so as to interfere with their detection, these deposits may be removed, and the field of the microscope cleared up, by the addition of a little acetic acid, if phosphates be present, and if urates, by a solution of soda or potassa (ten per cent. strength).

Clinical Importance.—This has been greatly over-estimated. Spermatozoids are frequently found in the urine of men who are in a state of perfect health. They are also met with in the urine first passed, *post coitum*, of both sexes, as well as

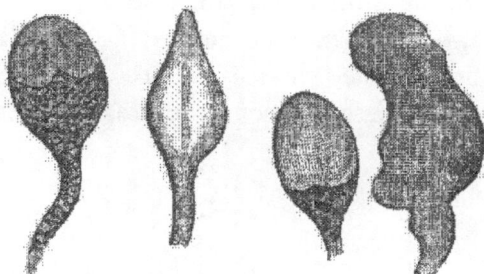

Fig. 32.—BODY AND UPPER PART OF THE TAIL OF A SPERMATOZOON.

✕ upwards of 3,000 diameters. (Beginning at left.) 1. spermatozoon containing much living germinal matter. 2. the same seen edgeways. 3. Spermatozoon containing comparatively little germinal matter. 4. Spermatozoon crushed, showing separate spherical particles of germinal matter. (Beale.)

after nocturnal emission in males. Unless they are met with more or less constantly in the urine, accompanied by other more important symptoms, they do not demand interference. Since their presence in the vaginal mucus, or upon the linen of a woman, is considered as being proof of sexual inter-course, or of an attempt at it, their detection becomes of great medico-legal importance in cases of suspected rape. A piece of the stained linen must be soaked in water, or, better, in an artificial serum, in a watch-glass or test-tube for an hour or two, and the sediment examined. Skilful and accurate ma-nipulation is required in such an investigation, and, up to a certain point, the higher the power used the better.

Fungi.

Many kinds of fungi (formerly supposed to be animal organisms, but now acknowledged to be vegetable in their nature) grow in the urine after it has been voided some time. The most important are (*a*) bacteria, (*b*) penicilium glaucum, (*c*) torula cerevisiæ, (*d*) sarcinæ. The latter only are apparently formed in the urine *before it is voided*.

1. *Bacteria.*—These may be defined as globular, oblong, rod-like or spirally-coiled masses of protoplasmic matter enclosed in a more or less distinct structureless substance. The

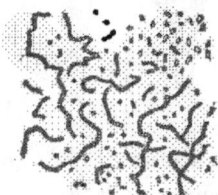

Fig. 33.—BACTERIA AND VIBRIONES.
From urine three days old. ×403. (Beale.)

smallest are not more than 1-30,000th of an inch in diameter, so that under the best microscopes they appear as little more than mere specks, and even the largest have a thickness of little more than 1-10,000th of an inch, though they may be very long in proportion, from 1-1000th to 1-3000th of an inch in length. Under a power of 200 diameters, when the field is strongly illuminated, may be seen myriads of very delicate linear bodies, exhibiting active motion. By a very high magnifying power these moving filaments are resolved into a series of granulated particles, arranged in a line. Many of them have two conditions—a still and an active state. In their still condition, however, they very generally exhibit that *Brownian* movement, which is common to almost all very finely divided solids suspended in a fluid. But this motion is merely *oscillatory*, and is readily distinguished from the rapid translation from place to place, which is effected by the really active *bacteria*. In the *vibriones*, as in all other forms of bacteria, the body does not rapidly change its form, but their joints are bent zig-zag-wise, and the rotation of the zig-zag upon its axis, gives rise to the optical illusion of a wriggling or serpentine motion. A corkscrew turned round, while its point rests against the finger, gives rise to just the same appearance. All the forms of putrefaction which are undergone by animal and vegetable matter are fermentations set up by bacteria of different kinds. Their vital actions are arrested at the freezing point; they thrive best in a temperature of about 74° F., but in most fluids they are killed by a temperature of 140° F. Ammonia and acetic acid arrest their motion without apparently otherwise affecting them; but one of the best reagents to render them conspicuous is iodine. It arrests their movements, colors their protoplasm yellow, without acting upon their sac or membrane. Salicylic acid, nitrite of amyl, as well as thymol, added to the urine, check the fermentation and putrefaction, by preventing the growth of these minute organisms. They swarm in urine beginning to putrefy, and their presence is certain proof that putrefactive changes have set in. The urine loses its transparency and deposits a cloudy sediment; its odor becomes offensive, and its reaction soon ammoniacal.

2. *Mould Fungus, or Penicilium Glaucum* (Fig. 34).—Urine, like other organic fluids, is liable to mildew; and it may mould from the growth of two distinct, though closely allied, vegetations. The first of these is the *penicilium glaucum*, or common blue-mould, which grows in vinegar and all albuminous fluids. It may be found in urine in three phases of its development; namely, as round and oval cells, or *sporules;* as a network of interlacing fibres, or

thallus; and as a downy pile of threads growing into the air, or *aerial fructification* (Fig. 34). This last phase is, however, not seen unless the urine has been kept several days. The *sporules* often appear in the urine a few hours after emission, and it is important to be familiar with their micro-chemical characters to avoid mistaking them for *blood discs.* Examine

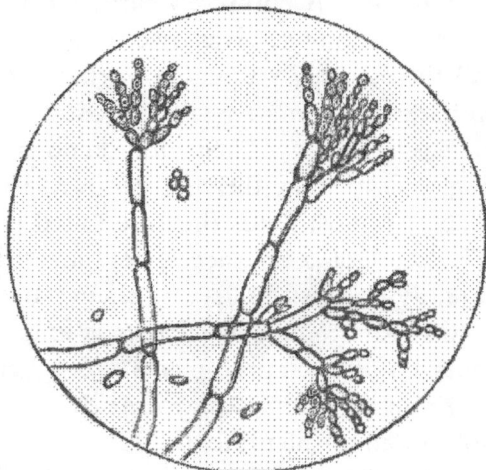

Fig. 34.—PENICILIUM GLAUCUM (MOULD FUNGUS). AERIAL FRUCTI-FICATION (Roberts).

with a magnifying power of 500 to 600 diameters. They are round or oval transparent bodies, varying in diameter from 1-2500th to 1-7000th of an inch (on the average about 1-3000th). (See *Blood*). The marks by which they are distinguished are : The great difference of size among the individual cells; the presence of a nucleus in the larger sporules; their tendency to assume an elongated or oval form; and (if set aside several days) the indications of budding and commencing formation of a thallus. Magenta or iodine stains their protoplasmic contents, but not their sacs. The *thallus* is produced by the elongation and gemmation of the sporules, and is composed of tubular cells placed end to end. This interlacement forms a fleecy cloud in the urine, which gradually rises to the surface, where it forms an islet or patch of white mould, from which springs the ascending stems of the aerial fructification. The latter consists of hollow filaments rising from the thallus, which divide at their extremities into two or three branches; these again sub-divide into a number of digitate projections, so as to form an irregular tuft or head. The digitate projections are filled with sporules, and eventually burst, giving exit to the sporules, which then fall into the urine below, and collect into an amorphous-looking deposit at the bottom of the vessel. Every *acid* urine (when it becomes ammoniacal, the further growth of the plant is arrested, and it soon perishes) forms a fitting nidus for the mould fungus, while albuminous urines are those in which the plant grows most luxuriantly.

3. *Yeast or Sugar Fungus* (Torula cerevisiæ), Fig. 35.— This fungus has precisely the same phases of development as the mould fungus; and in the phases of sporule and thallus, it is not easy to distinguish the one from the other; but the *aerial fructifications* of the two are wholly different. The yeast fungus (which grows luxuriantly in diabetic urine, exposed to a moderate temperature), instead of a tuft of branches, has a *spherical head*, not unlike that on the stem of an onion " going to seed." When the plant has attained its full fructification, the floating bed of thallus appears dusted over with a brown powder, which, under the microscope, is found to consist of these spherical heads full of sporules. The heads, when ripe, burst, and discharge their sporules, which sink to the bottom of the glass and form a white sediment like so much flour. The yeast fungus may grow to full fructification in urine in which the most delicate direct testing has failed to detect sugar; hence, until sugar is proved to be natural to normal urine, the sugar fungus cannot be received

as a proof of sugar in the urine.   Its spores may be distinguished from red blood corpuscles, which they resemble in

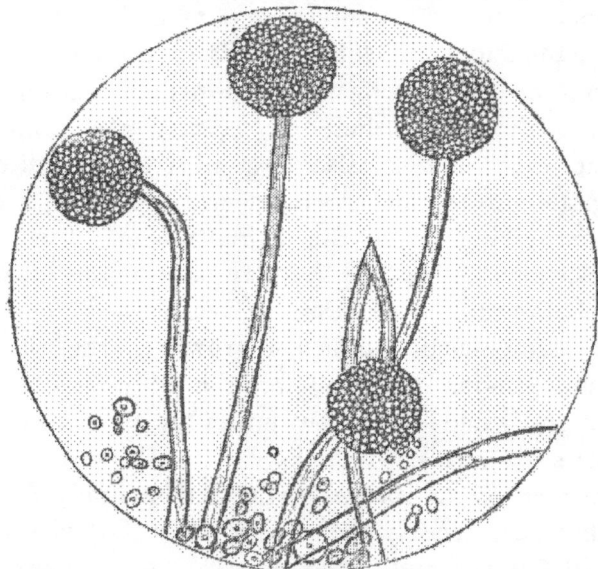

Fig. 35.—YEAST OR SUGAR FUNGUS (TORULA CEREVISIÆ).   AERIAL
FRUCTIFICATION.   (Roberts.)

size, by their generally oval shape, fatty lustre, and thin
granular contents in molecular movement, which the largest
ones show under a high power.

Fig. 36.—SARCINÆ IN URINE.   (Roberts.)

4. *Sarcinæ Ventriculi* (Fig. 36).—This is among the rarer forms of fungi found in the urine. They are square bodies, divided into secondary squares, which number 4, 16, 64, etc. and are like, but smaller than, the sarcinæ found in the vomited matters of persons suffering from stenosis of the pylorus. It exists both in acid and alkaline urine; and when found is generally associated with some disorder of the urinary organs, renal pains, dysuria, vesical catarrh, together with hypochondriacal and dyspeptic symptoms. Its clinical significance is obscure.

Selection of a Specimen for Examination.

If possible, obtain for examination a specimen from the whole quantity of urine passed in the twenty-four hours, which has been collected and mixed, since the specific gravity and reaction vary considerably at different times during the twenty-four hours. If, as is generally the case, we wish merely to ascertain the presence or absence of any particular substance, such, for example, as albumen, or sugar, the urine passed at any time of the day will in general suffice. But even in such cases, to render the observation of more value, it is better to employ the urine of digestion (*Urina cibi*), and that passed three or four hours after dinner is the best, since this invariably contains the greatest amount of any foreign substances, if such be present. Next to this it is better to take the urine first passed after rising in the morning (*urina sanguinis*). The urine may be received in a perfectly clean half-gallon magnesia jar, covered so as to exclude extraneous matters. *Examine, if possible, within three hours after being voided*, and in all cases before decomposition has begun. When received, a portion of the urine (about 4 or 5 ounces) should be poured into a tall, narrow, cylindrical glass vessel. A graduated test-tube, provided with a foot or base, is best, since it, at the same time, approximatively measures the bulk of any deposit that may fall.   The cylindrical vessels have the advantage over conical ones of having no sloping slides upon which the sediment may collect, and be thus prevented from falling to the bottom.   Cover the vessel carefully to exclude dust, and set it aside, whether there be a visible sediment or not.   Examine it microscopically after a few hours.   Re-examine it after the lapse of twelve hours, by which time any sediment that is likely to fall will have subsided.   While this portion is taking care of itself, divide the remaining urine into several parts, filtering them, if necessary, and examine them chemically according to the following tabulated scheme:

Systematic Qualitative Analysis of the Urine.

SECTION I.—GENERAL.

1. Color of the Urine.

1. Normal colors.
{ Pale urine; colorless to straw-yellow.
Ordinary urine; golden-yellow to amber.
Highly colored; reddish-yellow to brown.

2. Abnormal colors.

Essential; arising in the interior of the body.
{ Coloring matter of the blood.
Biliary Pigment, Urohæmatin, Uroerythrin, Indican.

Accidental; derived from without, and only passing through the organism.
Various coloring matters, as e. g.
{ Santonin, Saffron, Gamboge, Senna, Rhubarb, Aloes, Logwood, Gallic Acid, Creosote, etc.

2. Odor.

1. Essential.

Normal. { Sui generis.

Abnormal.
{ Whey-like. { Diabetes.
Ammoniacal { Decomposition.
Sulphuretted Hydrogen. { Decomposition.

2. Accidental, from odoriferous substances introduced into the organism.
Very much varied by
{ Asparagus, Garlic, Oil of Turpentine, Cubebs, Copaiba, Sandalwood Oil, etc.

3. Aspect.

Normal urine is always clear; clear urine is not necessarily normal.
The urine is turbid when voided. { *Pus, Mucus, Epithelia.*
The Urine is sedimentary. { *See Examination of Sediments.*

4. Reaction.

Normal urine has an *acid* reaction, due principally to the acid phosphate of the alkalies

A. Drop a very small slip of blue, and also one of red, litmus paper into the urine, and wait till they are completely saturated.
  { *Both* are red. { Acid.
  { *Both* are blue. { Alkaline. *See B.*
  { One is blue and the other red. { Neutral.

B. Dry the blue paper in the open air, or in a water-oven. (If yellow turmeric paper is used it will be *browned* by an alkali.)
  { Blue color persists after complete drying. { Fixed alkali. { *Potash Soda.*
  { Original red color is restored to the paper. { Volatile alkali. { *Ammonia.*

5. Specific Gravity.

1. Hold the urinometer-cylinder obliquely, *when filling it,* to avoid a foam.
2. Stand with the back to the source of light.
3. Hold the cylinder by the *top,* lightly, between the thumb and forefinger, letting it swing freely, by its own weight, in the perpendicular position, *so that the stem does not touch the sides.*
4. Read by the lower, sharp, convex edge of the liquid, the fluid being on a level with the eye.
5. If too turbid to read the scale, filter the urine before taking the specific gravity.

Normal. { Urina potus, $1002°-1015°$.
  { Urina sanguinis, $1015°-1020°$.
  { Urina cibi, $1020°-1030°$.

$1030°$ to $1065°$. { Test for *Sugar,* and excess of *Urea.*

Below $1015°$, persistently. { Test for *albumen.* Section 3.

$1005°$ to $1008°$, persistently. { *Diabetes insipidus.*

Table for Reducing the Indications of a Glass Urinometer to the Standard Temperature ($60°$ F.), when the Specific Gravity has been taken at a higher temperature. (*Bird*).

The urine must be of the temperature of the surrounding air, otherwise great errors may creep in.

Temperature	No. to be added to the Indication.	Temperature	No. to be added to the Indication.	Temperature	No. to be added to the Indication.
$60°$	0.00	$69°$	0.80	$78°$	1.70
$61°$	0.08	$70°$	0.90	$79°$	1.80
$62°$	0.16	$71°$	1.00	$80°$	1.90
$63°$	0.24	$72°$	1.10	$81°$	2.00
$64°$	0.32	$73°$	1.20	$82°$	2.10
$65°$	0.40	$74°$	1.30	$83°$	2.20
$66°$	0.50	$75°$	1.40	$84°$	2.30
$67°$	0.60	$76°$	1.50	$85°$	2.40
$68°$	0.70	$77°$	1.60	$86°$	2.50

6. Quantity.

From 40 to 60 fluid ounces (900 c.c. to 1500 c.c.) in the 24 hours.

7. Detection of Abnormal Substances in the Urine.

Divide the urine into several portions, in which search successively for the following substances:

Excess of Urea. See Section II.
Albumen. See Section III.
Sugar. See Section IV.
Bile. See Section V.
Blood. See Section VI.
Chyle. See Section VII.

8. Examination of the Sediments.

Chemically.—See page 13.   Microscopically.—See page 20.

SECTION II.—UREA, $(N\,H_2)_2 C\,O.$—EXCESS AND DEFICIENCY.

1.

Reaction of urine variable; color normal, specific gravity *over* 1030° (if excess be present).   Owing to its exceeding solubility, urea never forms a spontaneous deposit. A healthy man excretes from 300 to 600 grains in the 24 hours.

2. Test for Excess.

Take from the fresh-mixed urine of the 24 hours, about an inch in a test-tube.   Add to it one-third its volume of pure *colorless nitric acid,* and set the tube in water not warmer than 60° F.   It is best to work at the fixed temperature of 33° F., which is readily obtained by melting ice in water.

If crystals of *Nitrate of Urea* form immediately, or within a few moments, *urea is present in excess of the normal proportion.*

Nitrate of Urea shows flat, rhombic, or hexagonal plates, closely united to one another.   Colorless.

3. Test for Deficiency.

Take a sample of the same urine, evaporate over a water-bath to one-half its bulk at a low temperature, allow it to cool, add *nitric acid* as before, and set the tube in water at 60° F., or better, in water containing ice.

Crystals of Urea Nitrate do *not* form within a few moments, say five minutes.   *Urea is below the normal proportion.*

NOTE.—The quantity of urine passed in twenty-four hours must be taken into consideration, for if the patient is passing, instead of the normal 1500 c.c., say only 600 c.c., the urine should be diluted up to 1500 c.c. for test No. 2, and used *without evaporation* for test No. 3.   If, on the other hand, the patient is diabetic, and is passing 3000 c.c. per twenty-four hours, a given sample of his urine must be reduced to one-half by evaporation for test No. 2, and a portion of this reduced urine still further reduced one-half for test No. 3.

SECTION III.—ALBUMEN.

Normal Urine ought never to contain Albumen.

1.

a. The urine is turbid and sedimentary. ⟨ See 2.

b. The urine is clear and transparent. ⟨ See 2 *b.*

2.

Filter or decant.
{
a. The *sediment* must be examined under the **microscope for casts, epithelia, pus corpuscles, etc.**

b. The filtered urine is
{
c. Acid or Neutral. ⟨ See 3, Test.

d. Alkaline. ⟨ Neutralize with **nitric acid**, and see 3, Test.
}
}

3. Test.

G. Pour 1 c.c. of *nitric acid* into 10 c.c. (1 to 10) of urine in a test-tube.
{
a. Coagulation takes place. It is probably due to ⟨ *Albumen.* See *b.*

b. There is doubt; the liquid is only turbid, or there is but a slight precipitate. ⟨ See *H.*
}

H. Heat the urine.
{
c. The whole is redissolved. ⟨ *Uric Acid. Nitrate of Urea.*

d. The turbidity or precipitate remains. ⟨ *Albumen.* See *I.*
}

I. Add a small quantity of *alcohol.*
{
e. Turbidity disappears. ⟨ Resins. ⟨ Cubebs, copaiba, sandal-wood oil, turpentine, etc.

f. Turbidity remains. ⟨ *Albumen.*
}

4.

The following is a rough but useful *approximate quantity test:* Boil a given quantity of the urine in a graduated test-tube, with a drop or two of *acetic acid;* add *nitric acid,* and set aside for at least twelve hours. The precipitated albumen sinks, and forms a layer of varying thickness. The proportion of albumen is estimated by the depth of this layer, as compared with the height of the column of urine in the tube; and may be expressed in numbers as ½, ¼, 1-10. If too little albumen is present to form a layer, the proportion may be loosely expressed as a "cloudiness," or "opalescence." Greater accuracy is obtained by previously filtering the urine of urates, epithelium, or extraneous matter, which might unduly increase the bulk of the deposit on standing.

SECTION IV.—SUGAR, $(C_6 H_{12} O_6) + H_2 O$.—GLYCOSURIA.—DIA-
BETES MELLITUS.

1.

The essential features of the urine in diabetes are its *excessive quantity* and the *presence of sugar* (glucose). The daily *quantity* of the urine in diabetes oscillates usually between 128 and 230 fluid ounces (8 to 15 pints). It has been known to exceed 32 pints. When the excretion is considerable (exceeding 4 or 5 pints), the urine has a very *pale straw tint*, and a peculiarly bright aspect; a less quantity does not alter the natural general appearance. The proportion of sugar varies from 8 to 12 per cent.; the quantity excreted daily ranges from 15 to 25 ounces (may fall to an ounce or less; may rise to two pounds or more). The specific gravity of diabetic urine usually fluctuates a few degrees above, or below 1040°; it may rise to 1060°, or sink to 1015°. *The sp. gr. is no criterion of the amount of sugar present*, since coexistent *excess of urea* may cause a high density with little sugar, and *albuminuria* may cause comparatively low density with much sugar. If the quantity of sugar be large, a sweetish whey-like odor and taste are communicated to the urine. *Sediments are rarely observed in diabetic urine.*

2.

The cold urine, filtered or decanted, is *not* albuminous. { See 9.

3

The cold urine *contains albumen*. { See 7.

4.

The urine contains bile pigment, or is otherwise high colored. { See 8.

5.

The urine is acid or neutral (diabetic urine is usually acid). { See 9.

6.

The urine has an alkaline reaction due to *ammoniacal salts*. This interferes with the copper tests. So— { Boil some of the urine in a test-tube, with a small piece of *caustic soda* or potash; filter or decant. } See 9.

7.

Add a few drops of *acetic acid* to the urine; boil to coagulate the albumen, filter, neutralize the filtered urine with a little *sodium carbonate*, then test as per— } See 9.

8.

To *decolorize* the urine, put an ounce or two in an eight-ounce bottle, together with a tablespoonful of *animal charcoal*, and a small pinch of *sodium carbonate*. Shake well for five or ten minutes, and then filter. A perfectly colorless liquid will thus be obtained, and greatly facilitate the application of the copper tests. } See 9.

9. Test (Trommer's).

Add to a drachm of the suspected urine, in a test-tube, about five drops of a solution of *cupric sulphate* (10 grs. to the fluid ounce); then add of *caustic potash solution* (20 or 30 per cent. of strength) an *excess*—i.e., *until the precipitate* of the blue hydrated protoxide of copper *is completely dissolved.* Boil the *clear* blue solution (blue or black precipitate; no sugar). A *yellow* or red (suboxide of copper) precipitate. } *Sugar.*

NOTE.—Repeat the test once or twice with less of the *copper solution* each time. If the proper proportions have been used, the copper-test possesses both delicacy and certainty.

10. Differential Density Fermentation Test (Roberts.)

This test affords a simple and usually sufficiently accurate quantitative analysis for sugar. Proceed thus: (1) Place four fluid ounces of urine in a twelve-ounce bottle, with a lump of German yeast the size of a chestnut; cork loosely so that the carbonic acid gas may escape, or cover with a slip of glass, and set aside in a warm place, such as the mantel-piece, to ferment. (2) A companion bottle of the same size has put into it the same amount of the same sample of urine, *but no yeast is added,* and it is *tightly corked* and placed beside the fermenting vial. (3) In about 22 hours, *when fermentation has ceased,* the two vials are removed and placed in some cooler part of the room. (4) Two hours thereafter—that is, about 24 hours from the commencement of the experiment—the contents of both bottles are poured separately into cylindrical vessels, and the specific gravity of each taken with the urinometer. (5) The difference between the two specific gravities is thus ascertained, *and every degree of "density lost"* indicates *one grain of sugar per fluid ounce of the urine*— e.g., the sp. gr. of the *unfermented* urine = 1040°; of the *fermented* urine = 1020°; difference 20 degrees = 20 grains of sugar per fluid ounce. If the patient has passed, say, 80 ounces in the 24 hours, then $80 \times 20 = 1600$ grains of sugar excreted in the 24 hours.

———•◦•———

SECTION V.—BILIARY MATTERS.

1.

Color of the urine greenish-yellow to dark brown. Agitation produces a permanent yellow froth. White filtering (blotting) paper, or linen, stained yellow. } *Bile Pigments* present.

2. Gmelin's Test.

Spread a thin layer of the urine upon a white porcelain surface, and add to it a few drops of *nitroso-nitric* acid (nitric acid containing the lower oxides of nitrogen); that from the cell of a Grove or Bunsen battery answers perfectly. } The drop of acid is tinged with a rapidly varying play of colors (in the following order: green, blue, violet-red, and yellowish), which speedily disappear. The most characteristic and essential color is the green. *Bile Pigments.*

3. Pettenkofer's Test.

Place a small quantity of urine, which has been boiled and filtered, in a porcelain capsule (or watch-glass), and evaporate it *slowly* and gently over a water-bath, or spirit flame, until but a few drops remain; add to this a drop or two of a solution of pure *sulphuric acid* (1 part to 4 parts of distilled water), and then a drop of a saccharine solution (cane sugar 1 part, distilled water 4 parts).

A cherry-red color, rapidly deepening until it becomes a beautiful purple, or dark lake color, appears. Only the purple color is characteristic of the presence of the *Biliary Acids.*

NOTE.—Bile pigments have the property of adhering to sediments much more powerfully than other pigments, and may be detected in the sediment when not in the urine itself.

SECTION VI.—BLOOD.

A. On Cooling, the Urine Has a Blood-red Color.

1. It is transparent. *a.* Add a few drops of *Hydrochloric Acid.*
   1. The color becomes darker. *Coloring matters of the blood.* See B.
   2. The color becomes clearer. *Foreign coloring matters.* Rhubarb, etc.

2. It is slightly turbid. *b.* Allow it to rest until a precipitate forms, which examine. According to 3.

3. There is a red sediment. *c.* Observe, under the microscope, whether this precipitate is crystalline or amorphous; on heating it in a test-tube it becomes dissolved. *Urates. Uric acid.*

   d. The microscope shows blood corpuscles; the precipitate is not soluble by heat. *Blood.*

B. On Cooling, the Urine is Reddish-brown, "Smoky," or Ink-black.

4. There is no sediment, and the microscope shows no blood corpuscles. *Boil the urine* alone, or with a little *acetic acid.* *e.* A reddish-brown or dirty *coagulum* forms. *Coloring matters of the blood.* Hæmoglobin and products of its decomposition.

SECTION VII.—CHYLE.

1.

Reaction varies; sp. gr. varies; color milky, turbid, opaque (rose-colored, if blood be present). Consistence sometimes jelly-like.

2. Test.

a. Add to about *three inches* of urine in a test-tube, about an inch of *sulphuric ether.* Gently mix them. The urine assumes its natural color and transparency. { *Fat* is present, and has been dissolved by the Ether. See 3, *b.*

3.

b. Transfer a little of the Ether (floating on the urine), by means of a pipette, to a watch-glass (or slide), and allow it to evaporate. { A characteristic stain of fat remains on the glass or slide. *Chyle.*

Systematic Chemical Examination of Inorganic Urinary Deposits.

NOTE.—A portion of the specimen of urine that has been set aside in a cylindrical or conical vessel, has let fall a sediment. The first step consists in completely separating the deposit, which it is desired to examine, by *filtration.* The sediment remaining upon the filter, whatever be its character, must then be washed with a little distilled water. The various tests will be found under the following sections:

SECTION I.—URIC ACID ($C^5 H^4 N^4 O^3$).

1. General Appearance of Urine and Deposit.

Urine depositing uric acid has usually a rich yellow or orange color, and *invariably an acid reaction.* The uric acid crystals may form a film on the surface of the liquid, or lie scattered as brilliant brown specks on the sides of the glass, or subside into a dense red deposit (like cayenne pepper). In rare instances the crystals are so small as to require the microscope for their detection. Generally, however, the naked eye can identify uric acid with certainty, because no other *brown crystals* occur in urine as a spontaneous deposit. When the crystals are very minute, the deposit resembles the amorphous urates, but is *denser,* and sinks more rapidly. Uric acid is usually accompanied by a considerable sediment of urates. A healthy man excretes, on an average, about 7 or 8 grains of uric acid in the twenty-four hours.

2. Calcination at a Red Heat (on Platinum Spoon.)

No residue; gives off ammoniacal odor.
A light black porous coal, containing nitrogen, is sometimes left.

3. Solubility.

a. Insoluble in cold water (15,000 parts); slightly soluble in boiling water (1800 parts).

b. Insoluble in all *dilute* acids; alcohol; ether.

c. Soluble in caustic alkalies, carbonates of potash and soda, borax, acid phosphate of soda, from which, if a slight excess of an acid be added, crystals are re-precipitated.

d. Soluble in *strong* sulphuric acid, without undergoing decomposition.

4. Action of Alkalies.

Triturated with caustic alkalies, unctuous compounds are formed, and *ammonia is not set free.*

5. Action of Concentrated Nitric Acid.

Dissolves *with effervescence*. and forms a crystalline mass. { See 6.

Explanation of the reaction. { Uric acid is decomposed into } *Alloxan,* which forms the crystalline mass, and *Urea.*

Nitric acid is decomposed into } *Nitrous acid,* which, with *Urea,* give *Carbonic acid* and *Nitrogen.*

Causes of errors. } *Urates.* *Carbonates.* { Dissolve equally with effervescence, as well as calcined ammonio-magnesian phosphates. *(Beale.)*

6. Murexide Test.

Upon heating this mass, and slowly evaporating it to dryness, a *red residue* remains, which, treated with a few drops of *ammonia* (or caustic potash), becomes purple, or violet-red (murexide or purpurate of ammonia or potash). If potash has been used, the violet color disappears under heat. Caffein gives the same reaction.

7. Nitrate of Silver Test. (Schiff's.)

Dissolve traces of uric acid in *Carbonate of Soda.* With this solution touch, lightly, a paper upon which a drop of *Nitrate of Silver* has been allowed to spread. A dark spot (reduced nitrate of silver) is produced. Will detect from 1-1000 to 1-500,000 of *Uric acid.*

8. Approximate Estimation of the Quantity.

To 30 fluid ounces of urine add 3 fluid drachms of *Hydrochloric acid,* and set aside in a cool place, as a cellar, for 24 hours. At the end of that time the uric acid crystals, highly colored, will be found adhering to the sides and bottom of the vessel. Collect them on a filter. By using always the same quantities of urine and acid, a rough estimate may be made.

SECTION II.—AMORPHOUS URATES. (*Lithates.*)

1. Naked-Eye Appearance.

The "amorphous urates" occur usually in *acid* urine, of a high density (1027), as a bulky, loose pulverulent deposit, wholly devoid of *crystalisation.* Its color (varying extremely in tint and intensity, but always deeper than the urine from which it falls) may be fawn, orange, "brick-red," pink, or purplish. It usually sinks quickly and completely, except where the urine is *albuminous,* in which case the precipitate may continue a long while diffused throughout the urine, giving to it a milky appearance. It is the most common and least important of all the urinary sediments. Urine containing an excess of urates is never turbid when freshly passed; it is only when the urine *has cooled* that the peculiar muddiness is observed.

2. Shake, and Heat Some of the Urine in a Test-Tube.

a. The sediment dissolves, but reappears on cooling. } *Amorphous Urates.*
Urate of soda dissolves at about 100° F.
Urate of ammonia dissolves at about 200° F.

b. There is doubt; filter the *boiling* urine. } On cooling, the filtrate yields a deposit—*Urates.*
Filter, and apply *murexide* test to some of the deposit on the filter.

c. Add strong *Acetic acid* to some of the deposit. It dissolves, and recrystalizes as *Uric acid.* } *Urates.*

SECTION III.—OXALATE OF LIME.

1. General Appearance of the Deposit.

Urine depositing oxalate of lime is usually highly colored (dark amber hue), and acid (rarely neutral). The deposit itself is very scanty, colorless, and closely resembles a slight cloud of mucus. It is often conjoined with a deposit of uric acid and the amorphous urates. If freshly-passed urine be allowed to deposit oxalate of lime in a glass vessel, its sides will be traversed by very numerous fine lines, running in bands, transversely or obliquely, looking as if the glass were finely scratched. The sediment consists of two parts—a soft, pale-grey, mucus-like layer on the bottom of the vessel, and overlying this a snow-white denser layer, with an undulating but sharply defined surface. The form of the crystals of oxalate of lime is so characteristic, that there is seldom occasion to make use of *chemical* tests to determine them. They are too minute to be distinguished by the naked eye. Next to urates, oxalate of lime is the most common unorganized urinary sediment.

2. Solubility.

Soluble (readily) in { Mineral acids, *without effervesence;* also in acid phosphate of soda.

Insoluble in { Water, alcohol, ether, alkalies, vegetable acids. Distinguished from the phosphates by its *insolubility* in *Acetic acid.*

SECTION IV.—PHOSPHATES.

The phosphates are only separated from very feebly acid, or alkaline urine, and they are always deposited when the urine undergoes the alkaline fermentation.

1.

The urine is turbid or sedimentary. Divide into two portions, *a* and *b.* {
a. add a few drops of any acid. { It clears { *Earthy Phosphates.*
b. Filter or decant and { See 3.

2.

The urine has been recently passed, and is clear and limpid. Boil. { A precipitate is formed which is soluble in acids. } *Phosphates.*

3.

Add an excess of ammonia to the urine under investigation, agitate, and then allow to rest.
- *Earthy* phosphates are precipitated. Acids dissolve them. See 4, *A.*
- *Alkaline* phosphates remain in solution. See 6, *B.*

4. *A*—Earthy Phosphates.

Filter, or decant, throw upon the precipitate *Acetic acid q. s.* to completely dissolve it. { Neutralize with a few drops of *Ammonia*, then add a little *Ammonium chloride.* { Then add slowly and in *excess*, *Oxalate of Ammonia.* {

A precipitate falls. *Oxalate of Lime.* Examine it microscopically.

Ammonio-magnesian phosphate remains in solution; decant, or filter, and see 6, *B.*

5.

Add an excess of *Ammonia* to the decanted or filtered urine.
- 1. A precipitate. — *Ammonio-magnesian phosphate* (recognize under the microscope); soluble in acids, insoluble in water and alkaline solutions.
- 2. No precipitate. — Leave a little of the urine in a closed test-tube for twenty-four hours. If there be no precipitate, there is no *Phosphate of Magnesia* in the urine.

6. *B*—Alkaline Phosphates.

Divide the urine into two portions. Acidulate *a* with *Nitric acid.* { Add an equal volume of *Molybdate of Ammonia.* {
- A yellow precipitate. — *Phosphoric acid*, in the state of phospho-molybdate of ammonia, insoluble in acids, soluble in alkalies.
- No immediate precipitate. — There are only traces of phosphoric acid. Heat mixture to about 104° F.

From *b* drive off the ammonia by boiling, then separate the urine into two parts.
- Add *alcoholic* solution of *Bichloride* of *Platinum.* — A precipitate. { *Potassa*, in the state of chloro-platinate; beautiful yellow octahedral crystals.
- Add solution of *bi-meta-antimoniate of potassa.* — A granular precipitate. { *Soda.*

SECTION V.—CYSTINE ($C_3 H_7 N S O_3$).

1. General Appearance.

Very rare urinary sediment. A copious, light sediment (resembling fawn-colored urates), deposited from a urine feebly acid (or alkaline), of a yellowish-green color; oily appearance; peculiar sweetbriar odor (when fresh); very liable to decomposition (evolving ammonia and sulphureted hydrogen), and liable to be accompanied by ammonio-magnesian phosphates, mucus, and epithelia.

2. Solubility.

1. Soluble in
- Ammonia. { Beautiful six-sided *colorless* crystals reappearing when the volatile alkali has evaporated. They polarize light.
- Caustic alkalies. { Boiled with *Caustic Potash*, cystine yields ammonia and an inflammable gas.
- Mineral acids. { Forming crystalline compounds.
- Oxalic acid.

2. Insoluble in
- Acetic acid. { Precipitates it from its *alkaline* solutions, amorphously, or in imperfectly formed crystals,
- Ammonium carbonate. { Precipitates it from its *acid* solutions.
- Water, alcohol, vegetable acids (except oxalic).

3. Heated on Platinum Foil.

Burns in the air with a bluish-green flame, evolving thick white fumes, having a peculiar, offensive, acid, garlic-like odor, leaving a spongy charcoal, and staining the platinum a dark greenish-blue color, *which disappears under further heating.*

SECTION VI.—LEUCIN.

Very rare deposit in the Urine.

Pure Leucin is a white *non-crystallizable*, odorless, and tasteless organic, fatty-looking substance. If the urine be suspected to contain it, see 1.

1.

Evaporate slowly an ounce or two of the urine to the consistence of a syrup, and set aside to cool. Leucin, if present, appears as circular, oily-looking, dark-yellow, laminated discs, or spheres, *which float upon water.*

2. Solubility.

Soluble in
{
Boiling water, *very* ; alcohol, *sparingly.*
Strong acids.

Strong alkalies. { Dissolved in ammonia it remains unchanged, and yields larger spherules when the ammonia spontaneously evaporates.
}

Insoluble in { Ether. { This distinguishes it from *fatty matter*, which it so much resembles. From crystals of *Carbonate of Lime*, Leucin may be determined by *floating on water*—the lime crystals sink.

3. Test.

Put a small quantity of Leucin on a platinum spatula, add *Nitric acid*, evaporate carefully to dryness. Treat the residue with a few drops of *Caustic Soda* solution, which dissolves it. Gently heat to concentrate; an oily-looking drop is formed, which can readily be rolled about under the spatula, neither moistening nor adhering to it. (*Scherer.*)

NOTE.—As Leucin can rarely be had from the urine in sufficient quantity to apply this characteristic test, we have to rely entirely upon its *microscopic* characters.

SECTION VII.—TYROSIN ($C_9 H_{11} N O_3$).

Very rare deposit in the Urine.

1. General Appearance.

Greenish-yellow *crystalline* sediment, increasing considerably in bulk with slight evaporation of the urine.

2. To Obtain It.

In doubtful cases concentrate the urine, as recommended in the case of Leucin.

3. Test.

Boil the suspected deposit in an excess of water, and while boiling it add a few drops of a solution of *Nitrate of the Protoxide of Mercury* (nearly neutral); if Tyrosin be present, a red precipitate will go down, while the supernatant liquid will be colored rose or purple-red. (*Hoffmann.*)

4. Solubility.

Soluble in { Acids, alkalies, and boiling water.

Insoluble in { Ether, alcohol, and cold water.

5. Calcination.

Emits, when burned, a disagreeable burnt-horn odor, and does not sublime.

Examination of Urinary Calculi.

(After Witthaus.)

1. Heat a portion to redness on platinum foil:
   a. It is entirely volatile. See - - - - **2**
   b. A residue remains. See - - - - **5**

2. Moisten a portion with $N O_3 H$, evaporate nearly to dryness, add $(N H_4) H O$:
   a. A red color is produced. See - - - **3**
   b. No red color. See - - - - **4**

3. Treat a portion with $K H O$, without heating:
   a. An ammoniacal odor is observed. - *Ammonium urate.*
   b. No ammoniacal odor, - - - - *Uric acid.*

4. *a.* The $N O_3 H$ solution becomes yellow on evaporation; the yellow residue becomes reddish-yellow on the addition of $K H O$, and, on heating with $K H O$, violet-red, - - - - *Xanthin.*
   b. The $N O_3 H$ solution becomes dark brown on evaporation, - - - - - - *Cystine.*

5. Treat as in 2 :
   a. A red color is produced. See - - - **6**
   b. No red color. See - - - - **9**

6. Heat before the blow-pipe on platinum foil:
   a. Fuses, - - - - - - - - **7**
   b. Does not fuse, - - - - - - **8**

7. *a.* Colors the flame yellow, - - - *Sodium urate.*
   b. Colors the flame violet, - - *Potassium urate.*

8. The residue from 6:
   a. Dissolves in dilute H Cl with effervescence; the solution forms a white ppt. with ammonium oxalate, - - - - - - *Calcium urate.*
   b. Dissolves with slight effervescence in dilute $SO_4 H_2$; the solution neutralized with $(N H_4) H O$, gives a white precipitate with $P O_4 H Na_2$,
                                      Magnesium urate.

9. Treat as in 6:
   a. It fuses, - - - *Ammonio-magnesian phosphate.*
   b. It does not fuse, - - - - - 10

10. The residue from 6, when moistened with water, is:
   a. Alkaline, - - - - - - - 11
   b. Not alkaline, - - - *Tricalcic phosphate.*

11. The original substance dissolves in H Cl:
   a. With effervescence, - - - *Calcium carbonate.*
   b. Without effervescence, - - *Calcium oxalate.*

Systematic Microscopical Examination of Urinary Deposits.

REMOVAL OF THE DEPOSIT FROM THE VESSEL CONTAINING IT.

In order to remove the deposit from the lower part of the vessel in which it has subsided, the upper end of the *pipette* (a glass tube with both extremities open, one of which is much smaller than the other, and tapering), is firmly closed by the forefinger, and the tapering extremity carried down to the very bottom of the glass beneath, or at least in contact with,

the layer of sediment there formed.   The forefinger should
now be raised sufficiently to allow about an inch of the
calibre of the tube to be filled with the sediment, when the
upper aperture is to be again tightly closed, the tube withdrawn, about one-half of its contents allowed to flow out, and
the external surface wiped off with a piece of new linen or
muslin.   A certain quantity of the deposit, usually *a single
drop*, is now allowed to flow on the glass slide (which has
been scrupulously cleaned), or in a shallow cell, by gently
raising the forefinger from the top.   It is then covered with a
thin glass cover, and subjected to examination in the usual
way.   *Always examine with a lower power first, and then with
a higher one.*   Any excess of urine which may exude at the
margin of the cover, should be removed by bringing in contact with it a piece of filtering or blotting paper, or the ragged
edge of a piece of old muslin.

The removal of urinary deposits by means of a pipette is an
awkward and unsatisfactory mode of procedure, and several
ingenious contrivances have therefore been devised to facilitate
their collection and removal.   Small glass trays with glass
handles have been recommended; the tray is placed at the bottom of the urine-glass, and when the urine has deposited its
sediment, the tray is raised by means of the handle, and the
sediment examined.   In this way, however, the deposit is removed *en masse*, and successive layers cannot conveniently be
examined *seriatim.*   To obviate these difficulties, an improved
and exceedingly simple form of urine-glass, for the collection
and examination of deposits, has been contrived, by which
minute quantities of sediment can be withdrawn for the purpose of microscopic examination.   The arrangement consists
of a slight modification of Mohr's burette; this is shortened
and widened, forming a conical cylinder of sufficient capacity
to hold two ounces of urine, and width to allow a urinometer

to float freely. The sediment is drawn off by means of a glass jet inserted into a piece of india-rubber tubing, which is attached to the neck of the urine-glass. The flow is broken

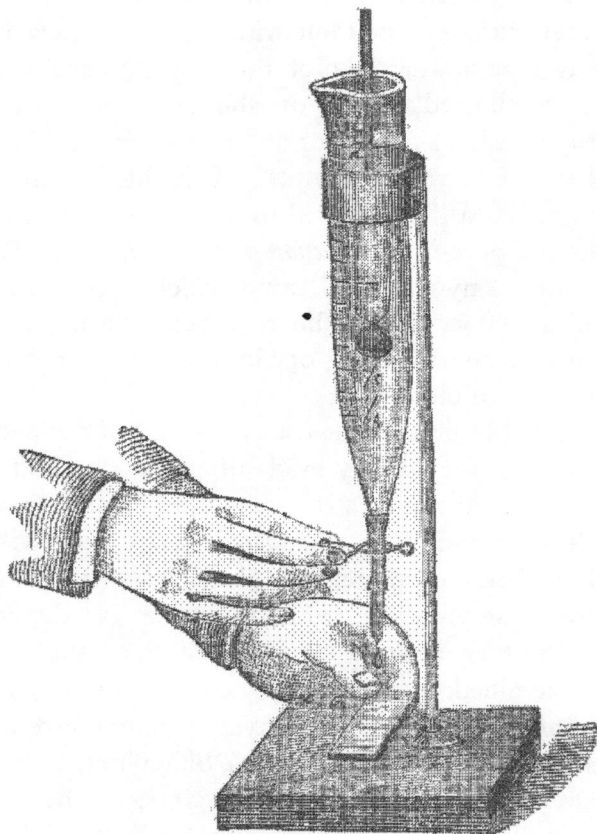

by a spring pinch-cock, which, by compressing the tubing between the neck of the urine-glass and the jet, effectually prevents leakage. On pressing the pinch-cock the compression is removable, and a flow occurs. Three or four pieces of india-rubber tubing fitted with glass jets are supplied with each instrument, and these when not in use are to be kept in a bottle filled with dilute hydrochloric acid, to prevent their becoming encrusted.   The whole instrument, mounted on a black stand with lacquered brass supports, is a neat as well as useful adjunct to the consulting-room table.

Thin shallow cells, made of glass, or gum dammar, or animalcule cages, present certain advantages over plain slides for examining urinary deposits, since in the former a stratum of fluid of any degree of thickness can be very readily obtained.   With a plain slide it is almost impossible not to *greatly modify* the microscopic appearance of the deposit by pushing the cover upon the stratum of urine between it and the slide.   Blood corpuscles and amorphous matter may, by this movement, be so aggregated together into cylindrical masses as to closely resemble tube-casts, and thus lead to serious error.   By using a shallow cell this source of error is avoided.

MAGNIFYING POWERS REQUIRED FOR THE EXAMINATION OF URINE.

Urinary sediments require to be examined with different magnifying powers.   The objectives which the author would recommend to the student are the two-third inch (giving a magnifying power, with different eye-pieces of from 75 to 150 diameters), and the one-fifth inch (250 to 700 diameters). If the above-mentioned objectives are of really excellent quality, additional magnifying power can be most economically obtained by means of an amplifier, or by using higher eye-pieces.   For urinary examinations, magnifying powers of from 200 to 600 diameters will, in general, answer all requirements.   The purchase, therefore, of a higher power objective than a one-fifth or one-sixth inch may be safely deferred by the student until he shall have become somewhat of an expert in microscopical manipulation.   Without any disparagement to the numberless excellent instruments in the market, within the reach of the student of limited means, the author can recommend, from personal experience in its use, the *New Working Microscope*, recently brought out by Mr. George Wale, an illustration of which will be found elsewhere.

THE MICRO-CHEMISTRY OF URINARY DEPOSITS.

In the investigation of those deposits which are prone to assume very various and widely different forms, such, for example, as uric acid, it will often be necessary to apply some simple chemical tests before the nature of the substance under examination can be positively determined.   By a little ingenuity and practice, the student may perform under the microscope all the chemical tests described in the foregoing *Systematic Chemical Examination of Inorganic Urinary Deposits*.   But *Nitric acid* and *Hydrochloric acid* should never be employed under the microscope, where it can be avoided, as the vapor from these acids rapidly corrodes the metallic mounting of the objective.   Whenever they are employed the objective should be promptly and carefully wiped with a piece of fine old linen, or a piece of soft glove leather.   In applying reagents, the pointed extremity of some blotting or filtering paper, or a thread, may be inserted between the slide and thin glass cover (or it may be applied closely against the latter), while a drop of the reagent is placed upon the other extremity of the paper at a short distance from the thin cover. The paper, absorbing the fluid by capillarity, establishes a current passing from the point upon which the reagent was

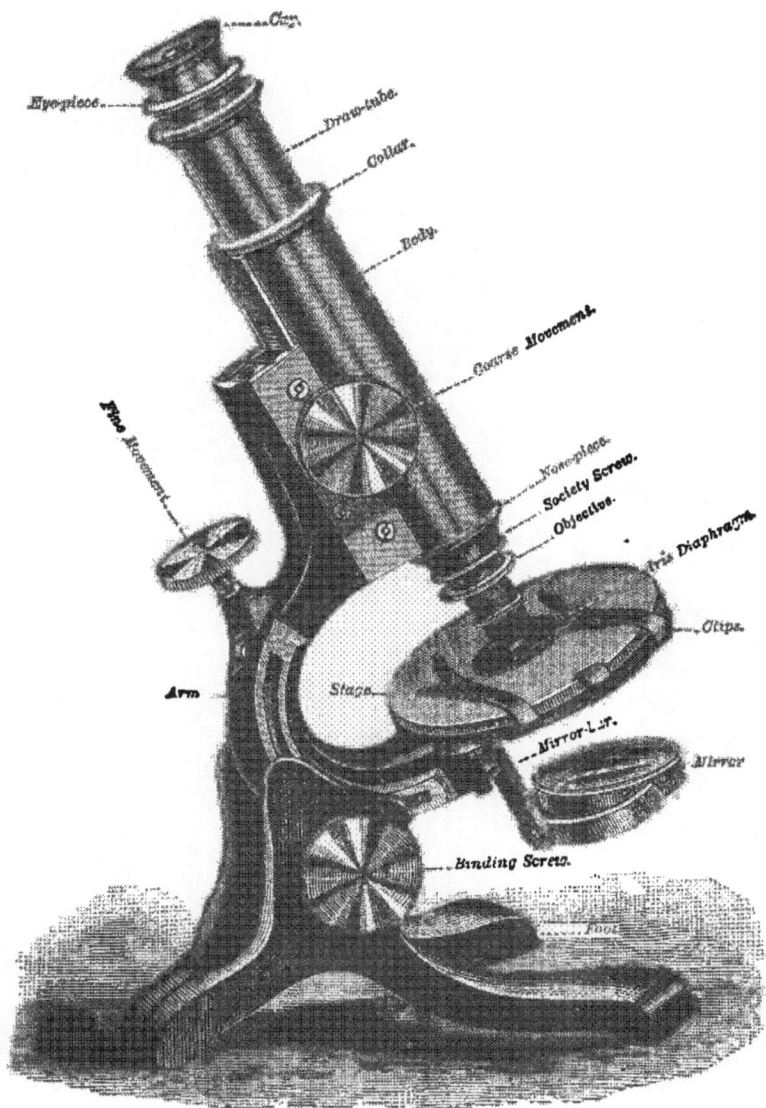

Eye-piece.

Cap.

Draw-tube.

Collar.

Body.

Coarse Movement.

Fine Movement.

Nose-piece.

Society Screw.

Objective.

Iris Diaphragm.

Clips.

Arm

Stage.

Mirror-Bar.

Mirror.

Binding Screw.

Foot.

NAMES OF THE DIFFERENT PARTS OF THE MICROSCOPE.

To aid the student in acquiring a familiarity with the microscope, we have inserted a cut in which the names of all the different parts are given in connection with each.

deposited, through the paper, to the thin glass. What occurs may then be observed, and the chemical reaction which ensues be investigated. A much more expeditious process consists in treating a small drop of the sediment upon a glass slide, with an excess of the reagent, then cover the whole with the thin glass, and examine the result under the microscope. A companion drop, not so treated, will show by comparison the changes which the reagent has induced. The following tables give the behavior of most urinary deposits when treated with *acetic acid.*

General Micro-Chemical Analysis of Urinary Sediments.

The simplest and, perhaps, best division of urinary deposits is into *unorganized* and *organized.*

A. UNORGANIZED SEDIMENTS.

(a) Crystallized.

1. *Acid Urine.*
   Uric Acid,
   Oxalate of Calcium,
   Calcium Phosphate (Stellar),
   Cystine,
   Tyrosine.

2. *Alkaline Urine.*
   Ammonium Urate,
   Triple Phosphate,
   Calcium Phosphate,
   Magnesium Phosphate.
   (Very rare.)

(b) Amorphous.

Urate of Sodium and Potassium,
Fats.

Calcium Carbonate,
Calcium Phosphate.

SECTION I.—NON-ORGANIZED BODIES.

1. Distinctly Crystalline Bodies.

Very large *transparent* crystals with sharply defined edges, generally isolated. Typical form, a triangular prism with beveled edges. Occurs in *alkaline* urine only. Polarizes light beautifully.	*Soluble* in Acetic acid.	*Ammonio-magnesian (or Triple) Phosphate.*
Large crystals, often grouped, *always colored* (in red, yellow, or brown); typical forms, a four-sided rhomb or hexagonal plate; surface often fissured; outlines very dark. Occurs in *acid* urine only. (Use a power of 100 or 200 diameters.)	*Insoluble* in Acetic acid.	*Uric acid.*

Large crystals, *colored*, closely resembling ammonio-magnesian phosphate in form, may occur in needles, either separate or combined in stellate groups. Occurs only in *acid* urine. *Very rare deposit.* (Use a power of 200 diameters.) } *Insoluble* in Acetic acid } { *Hippuric acid.*

Very small crystals (less than 1-1000th of an inch in diameter, usually), isolated; octohedral form (more rarely dumb-bells); very transparent, or of a faint greenish tint; very refractive; sharp edges; require a power of 400 to 600 diameters to show them well. } *Insoluble* in Acetic acid. } { Crystals of *Oxalate of Lime.*

Irregular, opaque, globular masses, or spherules, with spiny projections (either straight or curved), "hedge-hog" crystals—or prismatic crystals *arranged in star-like* masses. Comparatively rare spontaneous deposit in the *crystalline* form. Found in *acid* and *neutral* urine. } *Urate of Soda.*

Crystalline rods, either separate or in stellate groups (rosettes), or sheaf-like bundles. Some of the crystals club, wedge, or bottle shaped, and abundantly marked with lines of secondary crystallization. Frequently associated with oxalate of lime. Not a very common deposit. Found in pale, faintly *acid* urine, with a tendency to alkaline fermentation. } *Soluble* in Acetic acid. } { *Crystalline* Calcium. "*Stellar*" *Phosphate.*

Regular, colorless, hexagonal tablets; various sizes; united by their flat surfaces, and overlapping one another; may have an iridescent mother-of-pearl lustre; surfaces often marked with lines of secondary crystallization; or (being dimorphous) square prisms, singly or stellate; strongly refracting light. *A very rare deposit.* } *Insoluble* in Acetic acid. (*Soluble* in Ammonia. This differentiates it from Uric acid.) } { *Cystine.*

More or less yellowish-tinged, highly-refractive spheres, having the appearance of fat globules, with sharp contours, and, with good light, showing radii and delicate concentric lines. Found only in grave destructive diseases of the liver. } *Insoluble* in Ether. (Distinguishes it from Fat.) } { *Leucin.*

Very fine short silky acicular prisms, or needles, arranged in beautiful bundles, tufts, or "sheaf-like" collections or spiculated balls. Usually of a deep yellow color from absorbed bile pigments. Found associated with and under the same circumstances as the preceding body. } *Insoluble* in moderately strong Acetic acid. } { *Tyrosin.*

2. Amorphous Bodies.

Granules, irregular, opaque; or spherules, from which project spiny crystals, straight or curved; or globules, opaque, isolated, or united in a mass like frogs' eggs.	Slowly *soluble* in Acetic acid, after a short time giving rise to colorless tablets of Uric acid.	*Urates.*
Granules, roundish or oval, with dark outlines, isolated, or three or four united in a star-like form, or in beads, etc. Granules very pale, much smaller, very transparent, and difficult to perceive; always united by irregular punctated patches (the most common appearance).	*Soluble* in Acetic acid.	*Calcium Phosphate.*

SECTION II.—ORGANIZED BODIES.

1. Mucus.   2. Epithelium (from various parts of the genito-urinary tract).   3. Renal Tube Casts.   4. Blood.   5. Pus.
6. Spermatozoa.   7. Fungi.   8. Entozoa.

[NOTE.—Urine voided turbid will, in the majority of cases, owe its turbidity to one or more of the organized deposits.]

1. Cellular, Round, or Oval Bodies.

Circular discs or globules; *non-nucleated*; 1-3500th of an inch in diameter; separate; edges smooth or dentated; transparent or of a faintly yellowish color; sometimes presenting a central depression, and, if seen in profile, bi-concave. (Use high power.)	Swell in weak *Acetic acid*, or shrink and present a "raspberry" aspect. *Not colored* by carmine.	*Red Blood Corpuscles.*
Globules, round or oval; *nucleated*; 1-2000th to 1-1400th of an inch in diameter; slightly defined outlines; single, double, or multiple nuclei; grayish-white or granular contents; isolated, or united in masses of polygonal cells (use 300 to 400 diameters).	Rendered pale by *Acetic acid*, which causes two or three nuclei to appear within them. *Colored* by carmine.	*White Blood Corpuscles or Leucocytes.*
Globules, round, elongated, or oval; very small, varying in diameter from 1-7000th to 1-2500th (average 1-3000) of an inch; transparent, very refractive; larger ones nucleated, or have vacuoles, and sometimes warty expansions. Germinate if set aside several days. United in chains of three and four, or more, or isolated. (Examine with magnifying power of 500 to 600 diameters.)	Unchanged by *Acetic acid*. Uncolored by carmine. Their protoplasmic cell-contents, but not their sacs, stained brownish-yellow by *Iodine-water*.	*Spores of Fungi (Penicillium Glaucum, etc.)*

Corpuscles very small (length about 1-3000th of an inch); very refractive; of a peculiar bluish tint, and fatty lustre; furnished with a very delicate, long, (1-250th of an inch) tail-like filament; general appearance of a minute tad-pole, with greatly elongated tail. (Use 400 to 600 diameters.) } Unchanged by the reagents above mentioned. } *Spermatozoa*

2. Form Variable; Size Greater than the Preceding.

Round, oval, lamellar, cylindrical, fusiform, caudate, or irregular bodies: furnished generally with one or more nuclei, with granular contents. } Rendered pale by *Acetic acid*, which brings out their nuclei very distinctly. *Colored* (the nuclei especially) by *carmine*. } *Epithelium* from various parts of the Genito-urinary tract.

3. Cylindrical.

A. Voluminous, of greater or less length (rarely exceeding the 1-50th of an inch); variable aspect; pale or hyaline, granular or covered with epithelium: sometimes distinctly, sometimes indistinctly outlined; generally round or club-shaped extremities. } *Renal Tube Casts.*

a. Very pale or transparent Amorphous cylinders. {
With badly defined margins, often twisted or varicose, branching and subdividing. } *Uncolored* by carmine. } *Mucus Casts.*

With clear, well-defined margins, sometimes intersected by fractures. } *Colored* by carmine. } *Hyaline or "waxy casts."*
}

b. More or less dark, epithelial or granular cylinders. (Use power of 200 to 400 diameters.) {
No line of contour: epithelial cells united into a cylinder. Never very narrow. } *Epithelial or granular casts.*

A more or less distinct line of contour: fundamental substance finely granular; studded with blood corpuscles. } *Fibrinous blood casts.*
}

Any of the above may undergo fatty degeneration.

B. Very short; very small (1-1000th to 1-3000th of an inch in *length*) transparent bodies; sometimes motionless, but generally exhibiting active vibratory movements; or two or more joined end to end with a spiral motion. (Requires a power of 500 diameters, and upwards, to show them well.) } Unchanged by *Acetic acid*, except the arrest of motion. Their protoplasm colored yellow by *Iodine-water*. } *Bacteria.*

4. Filamentous, or Fibrillary.

Very thin : more or less ramified or interlacing.	*Acetic acid* does not change them.	*Thallus of Fungi.*
	Acetic acid renders them pale, causes their fibrillary aspect to disappear, and gives rise to a swollen, transparent, amorphous mass.	*Fibrin.*
	Acetic acid renders them more distinct, and gives them a punctated or striated appearance.	*Mucus.*

5. Square.

Square bodies, subdivided into secondary squares, which number 2, 4, 8, etc., and when collected in the form of cubes, very much resemble bales of goods.	*Sarcina.*

Form of Recording Urinary Examinations.

(To be printed on a full letter sheet.)

Examination of Urine.

For   at the request of

Dr.

Physical and Chemical Characters.

Total Quantity in twenty-four hours,

Color and Appearance,

Odor,

Reaction,

Specific Gravity,

Albumen,

Sugar,

Quantity and General Appearance of the Deposit,

Microscopical Appearance.

Crystals,

Anatomical Elements,

Casts,

Other Morphological Elements.

Remarks.

INDEX.

Acid fermentation of the urine, 16, 36.
Albuminuria, 29 to 31.
  tests for, 105.
Alkaline fermentation of the urine, 16, 36.
Alkalinity of the urine, 15.
  significance of, 17, 18.
  artificial production of, 16.
  produced by diseases, 16.
Ammonium carbonate, 16, 17.
Ammoniacal urine, 17, 18.
Ammonium urate, 17, 36, 55, 115.
Ammonio-magnesian (or triple) phosphate, 17, 36, 37, 63, 64.
  microscopic appearances of, 64, 65, 123.
  as calculi, 116.
Apparatus for quantitative estimation of urea, 45.
  collecting urinary sediments, 118.
Bacteria, 36, 96, 126.
Biliary acids in the urine, 108.
Bile in the urine, 34.
  tests for, 107, 108.
Bilirubin, 34.
Bird's specific gravity and temperature corrections, 103.
Blood in the urine, 87, 90.
  clinical significance of, 90.
  chemical tests for, 108.
  microscopic appearances of, 88, 89, 125.
  origin and occurrence of, 87.
Bouchardat's specific gravity and temperature corrections, 21.
Calculi, urinary, 115.
  examination of, 115.
Casts in the urine, 78, 87, 126.
  blood, 81, 82, 126.
  classification of, 79.
  clinical significance of, 86.
  collection and examination of, 78.
  diagnosis of, 85, 86.
  epithelial, 80, 81, 126.
  fatty or oil, 81, 82, 126.
  granular, 80, 81, 126.
  hyaline, waxy, 82, 83, 84, 126.
  mucus, 73, 84, 85, 126.
  of the seminal tubules, 85.
  origin or mode of formation of, 79, 80.

Chylous urine, 11.
  test for, 108, 109.
Color of the urine, 12, 102.
  modified by bile pigments, 14, 102.
  modified by indican, 14, 102.
  modified by medicines, 14, 102.
  modified by poisons, 14, 102.
Composition of the urine, 7.
Consistency of the urine, 11.
Cystin, 68.
  as calculi, 115.
  clinical significance of, 70.
  general appearance of deposit of, 113.
  microscopic appearances of, 67, 70, 124.
  tests for, 115.
Cystitis, 18, 92.
Deposits in the urine, 35.
  apparatus for collection of, 117 to 119.
  classification of, 123.
  definition of, 35.
  examination of, 109 to 114.
  from changes in reaction, 35, 36.
  from standing, 10.
  micro-chemical analysis of, 120, 123 to 127.
  non-organized, 123 to 125.
  organized, 125, 127.
Diabetes mellitus, 31, 32, 106.
Diabetic sugar, 32.
Epithelium, 74 to 78, 126.
Extractive matters in the urine, 33, 34.
  test for, 34.
Extraneous substances found in the urine, 38, 39.
Fluorescence of the urine, 14.
Form for recording urinary examinations, 128.
Froth on the urine, 11.
Fungi in the urine, 96 to 101, 126.
Gases in the urine, 8.
Glycosuria, 31, 106.
Gmelin's test for bile pigments, 107.
Hæmaturia, 87 to 90.
Hæmoglobin, 12.
Hæser's formula, 23.
Highly-colored urine, 12.

Hippuric acid, 56.
 microscopic appearances of, 57, 124.
Indican in the urine, 14, 102.
Leucin in the urine, 67, 68, 113, 124.
 tests for, 114.
Magnifying powers for urinary examina-
 tions, 119.
Micro-chemistry of urinary deposits, 120,
 123.
Microscope, 121.
Mould fungus, 97 to 99.
Mucus, 72 to 74, 127.
Mucus casts, 73, 84, 85.
Murexide test for uric acid, 110.
Odor of the urine, 11, 102.
 how modified, 12, 102.
Opacity of the urine, 10.
Oxalate of lime, 57 to 61.
 clinical significance of, 58.
 general appearance of deposit, 111.
 microscopical appearances of, 59, 124.
 solubility of, 111.
Pale urine, 12, 13.
Penicilium glaucum, 36, 97, 98, 99, 125
Pettenkofer's test for biliary acids, 108.
Phosphates, 61 to 67.
 clinical significance of, 66.
 tests for the, 110, 111.
Phosphate of lime, amorphous, 17, 37, 62,
 115, 125.
 crystalline, 62, 63.
 microscopic appearances, 62, 124.
Pigments of the urine, 12.
 affinity for urates and uric acid, 14.
Purpurin, 13.
Pus in the urine, 90 to 93.
 occurrence and recognition of, 90.
 clinical significance of, 92.
 microscopical appearances of, 91, 92,
 125.
Pyuria, 90 to 93.
Quantity of the urine, 24, 104.
 modified by cutaneous exhalations, 26.
 modified by diseases and remedies, 27.
 modified by drink, 24.
 modified by food, 25.
 modified by pulmonary exhalations, 26.
 modified by sex and age, 26.
 modified by sleep, 27.
 modified by stools, 26.
Reaction of the urine, 15 to 18, 103.
 modified by diseases, 15.
 modified by food, 15.
 modified by medicines, 15.
 modified by standing, 16.
 significance of, 18.
 tests for, 103.
Renal function, 8.
Renal tube-casts, 78 to 87, 126.
Robert's fermentation test, 107.
Sarcina in the urine, 100, 101, 126.
Scherer's test for leucin, 114.

Schift's test for uric acid, 110.
Sediments in the urine (see deposits).
Selection of a specimen for examination,
 101, 102.
Specific gravity, 18 to 23.
 determination of amount of solids by
 the, 23.
 method of finding the, 21, 102.
 modified by diseases, 19.
 table for correcting for temperature,
 21, 103.
Spermatozoa in the urine, 93 to 95, 126.
 clinical significance of, 95.
Sugar in the urine, 31, 32, 106.
 tests for, 106.
Systematic qualitative analysis of urine,
 102 to 127.
Table for specific gravity and temperature
 corrections. 21, 103.
Test for abnormal substances in the urine,
 104.
 albumen, 105.
 alkaline phosphates, 112.
 amorphous urates, 111.
 bile acids, 108.
 bile pigments, 107.
 blood, 108.
 chyle, 109.
 cystine, 113.
 deficiency of urea, 104.
 earthy phosphates, 112.
 excess of urea, 104.
 leucin, 114.
 oxalate of lime, 111.
Torula cerevisiæ, 36, 99, 100.
Transparency of the urine, 10.
Trapp's formula, 23.
Trommer's test for sugar, 107.
Tyrosin, 67, 68, 124.
 test for, 114.
Test for sugar in the urine, 106.
 tyrosin, 114.
 uric acid, 110.
Urates, amorphous, 51, 110, 125.
 amorphous, test for, 111.
 amorphous, colored by pigments, 14.
 crystalline, 53.
 precipitated by acid fermentation, 16,
 36.
 precipitated by cold, 10.
 precipitated by pathological conditions, 52.
 precipitated by physiological conditions, 52.
 microscopical appearances of, 52.
Urate of ammonium, 17, 36, 55, 111, 115.
 calcium, 116.
 magnesium, 116.
 potassium, 115.
 sodium, 53, 54, 111, 115, 124.
Urea, 38 to 46.
 decomposition of, 16, 17.

Urea modified by diet, 42.
  nitrate of, 40.
  oxalate of, 41.
  quantitative test for, 43 to 46.
  test for deficiency of, 104.
  test for excess of, 104.
Uric acid, 46 to 50, 109.
  action of alkalies on, 110.
  approximate estimation of, 110.
  artificial production of, 50.
Uric acid calculi, 115.
  clinical significance of, 47.
  colored by pigments, 14, 48.
  common forms of, 49.
  general appearance of deposit of, 109.
  microscopical appearance of, 49, 123.
  murexide test for, 110.
  nitrate of silver test for, 110.
  rare forms of, 50.
  solubility of, 109.
  test by calcination, 109.
Urine, acidity of, 7, 8, 15.
  acid fermentation of, 16.
  ammoniacal, 17.
  black, 14.
  blue, 14.
  color of, 12, 102.
  consistency of, 11.

Urine, extraneous substances in the, 38.
  fluorescence, 14.
  gases in, 8.
  glass for collecting deposits, 118.
  green, 14.
  high-colored, 12, 13.
  odor of, 11, 102.
  opacity of, 10.
  pale, 12, 13.
  quantity of, 24, 102.
  reaction of, 15 to 18, 103.
  specific gravity of, 18 to 24, 102.
  turbidity of, 10, 72.
  when to examine, 37, 101.
Urinary analysis set and stand (see frontis-
    piece).
Urinary pigments, 12, 102.
Urinometer, 20.
Urobilin, 12, 13.
Urochrome, 12.
Uroerythrin, 13, 102.
Urohæmatin, 13, 102.
Variations, physiological and patholo-
    gical, 9.
Xanthin, 71.
  microscopical appearances, 71.
  calculi, 115.
Yeast fungus, 99, 100.

www.ingramcontent.com/pod-product-compliance
Lightning Source LLC
Chambersburg PA
CBHW021936190326
41519CB00009B/1029